Kasaye Bahiru Tola, Girma Salale Geleta
Practical Chemistry

Also of Interest

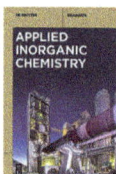

Applied Inorganic Chemistry
Volume 1–3
Rainer Pöttgen, Thomas Jüstel and Cristian A. Strassert (Eds.), 2022
Set-ISBN 978-3-11-074233-6

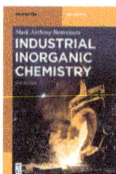

Industrial Inorganic Chemistry.
2ED
Mark Anthony Benvenuto, 2024
ISBN 978-3-11-132944-4
e-ISBN 978-3-11-132951-2

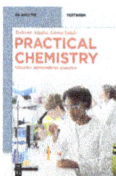

Practical Chemistry, Volume 1.
Instrumental Analysis
Teshome Adugna, Girma Salale, 2024
ISBN 978-3-11-157504-9
e-ISBN 978-3-11-157563-6

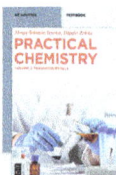

Practical Chemistry, Volume 2.
Transition Metals
Mesay Solomon Tesema, Digafie Zeleke, 2024
ISBN 978-3-11-157384-7
e-ISBN 978-3-11-157434-9

Kasaye Bahiru Tola, Girma Salale Geleta

Practical Chemistry

Instrumental Analysis and Quantitative Analytical Chemistry

Volume 3

DE GRUYTER

Authors
Kasaye Bahiru Tola
Department of Chemistry
College of Natural Sciences
Salale University
P.O. Box 245
Oromia, Fiche, Ethiopia
kasayebaharu@gmail.com

Dr. Girma Salale Geleta
Department of Chemistry
College of Natural Sciences
Salale University
P.O. Box 245
Oromia, Fiche, Ethiopia
ggirma245@gmail.com

ISBN 978-3-11-170221-6
e-ISBN (PDF) 978-3-11-170226-1
e-ISBN (EPUB) 978-3-11-170234-6

Library of Congress Control Number: 2024952071

Bibliographic information published by the Deutsche Nationalbibliothek
The Deutsche Nationalbibliothek lists this publication in the Deutsche Nationalbibliografie;
detailed bibliographic data are available on the internet at http://dnb.dnb.de.

© 2025 Walter de Gruyter GmbH, Berlin/Boston, Genthiner Straße 13, 10785 Berlin
Cover image: shironosov/iStock/Getty Images Plus
Typesetting: Integra Software Services Pvt. Ltd.

www.degruyter.com
Questions about General Product Safety Regulation:
productsafety@degruyterbrill.com

Preface

This manual is provided for use in a semester's Practical Instrumental Analysis II course. It is compiled based on the modularized curriculum for BSc in chemistry. The experiments in this handbook have been selected for their practical manipulations and educational usefulness. The primary goal in creating this handbook is to enhance some of the older experiments by adding new ones and modifying others, which will help better connect the "theoretical course" that is being taught concurrently with the practical course.

Along with novel experiments, this handbook includes a report writing form, additional safety instructions where needed, especially for students, and instructions on how to make them convenient for technicians and students. Students are required to work in pairs or more at each setup location in the laboratory. No member of any group can just stand by and do nothing. Cleaning the workspace, glassware, and equipment at the conclusion of each experiment is crucial and should be expected of the students. Maintaining discipline and following safety procedures during every experiment help prevent accidents and are a must when working in a laboratory.

This laboratory course aims to teach students about what happens in actual chemistry labs by connecting the theoretical and practical aspects of the lab. Students are expected to become familiar with the steps involved in each experiment before visiting the lab. They should arrive with a lab handbook and flowchart, and should also turn in their lab reports on time for each experiment.

https://doi.org/10.1515/9783111702261-202

Contents

Laboratory safety rules guidelines

Please follow the following laboratory safety rule when you perform an experimental activity in the laboratory:

– In the lab, never smoke, chew gum, eat, or drink any liquids, including water.
– With appropriate protection, the skin is less vulnerable to chemical exposure. It is therefore essential to wear a lab coat and full-leg pants when dealing with chemicals in the lab.
– Shoes: Open-toed shoes, such as flip-flops and sandals, are not allowed since they expose the feet to chemicals and sharp items. Long hair must be secured tightly since hair and loose clothing are easily combustible.
– Before entering the lab, ensure the supervisor or instructor is present.
– The supervisor must be notified right away of any accidents or injuries.
– Chemicals should normally be stored in areas with designated access. All lab reagents must have their caps changed as soon as possible. Chemicals should be transferred or dispersed carefully. Spills must be cleaned up right away and thoroughly.
– If any chemical substance spills onto the skin, wash with lots of water and let your teacher know.
– The students are responsible for reading labels and using the right chemical at the required concentration. Certain reagents can only be distributed by the instructor; these will be mentioned in the lab introduction.
– Glassware that is chipped or fractured should not be used. Follow your teacher's instructions for proper disposal.
– Lab safety depends on keeping a clean workspace and cleaning up after yourself. The weighing areas and bench tops need to be cleaned. All shared equipment should be cleaned. Expensive equipment can become worthless due to improper cleaning.
– Only with permission from the instructor or in accordance with any guidelines provided in the lab, the equipment can be used.
– It is the obligation of the student to dispose of waste chemicals properly. Ask the supervisor if you have any questions.
– Before you visit the lab, read the full experiment and finish any pre-assignments.
– It is necessary to notify the instructor of any rare health conditions, such as chemical allergic response, as soon as feasible. Constantly exercise common sense in the lab. Ensure to ask the teacher any questions you have before continuing.

https://doi.org/10.1515/9783111702261-204

Guidelines for writing laboratory reports

The important information that should be included in describing preparative types of experiments is the following in addition to the date on which the experiment is carried out (date should be written at the top right side above the title).

Title: This is the main topic of the experiment.

Experiment number: This is the experimental number you have carried out.

Objective: This is the aim of experiment you have carried out and it should start with the word "To."

Introduction (theory): Usually, a lab report's introduction begins with a description of the goal or findings of the experiment. Give a succinct explanation of why this could be interesting. The chemicals to be researched, the methods to be employed, and their interpretation should all be introduced. It would be acceptable, for example, to talk about how IR, NMR, and ultraviolet–visible (UV–Vis) spectrophotometers can be used to characterize chemicals that you would synthesize in the lab. Generally speaking, the questions and concerns that will form the main body of your Discussion section should be outlined in the Introduction.

Apparatus and chemicals: These are all of the tools and materials you used to conduct the experiment.

Procedure: This is a step-by-step guide to conduct an experiment and should be written in a passive voice.

Result (observation)
This part should be narrative, outlining your actions and the outcomes you achieved. It should explain your process, not the one in the handbook. It should contain the yields of the products you obtained (in both grams and percent) as well as the quantity of reagents you used. Rather than conventional procedures (assuming that we all know how to use a balance, a glove box, and an IR spectrometer, for example), physical measuring techniques should also be included. It is highly recommended that chemical equations be balanced. Labels for the atoms or groups of interest should be displayed on diagrams depicting the complexes' proposed and actual (if different) structures. These labels can subsequently be referred to in the text and used in the data table for assignment purposes. Whenever feasible, numerical data should be displayed in tables as opposed to within the text.

Then, as "absorbance readings were obtained over the wavelength range 280–340 nm and the data are given in Table 1" the text might make reference to the table. Any issues with gathering data should be described. Provide the actual data whenever

https://doi.org/10.1515/9783111702261-001

you can, together with any pertinent equations that demonstrate how the quantities of interest were determined. [Note in grammar: data is a plural noun.]

Discussion

Although not necessary, separate Results and Discussion sections are advised. Your analysis of the findings should come first in the Discussion section. Have you created what you desired? How are you aware? Talk about the reasons why the predicted product might not have formed and whether the real product can be identified. If something went wrong, talk about it. For instance, you may conduct comparisons between the spectral data of the products and the starting components. Impact and Connection Significant data and error analyses are needed for Isomers labs, and these can be included in the Results (if you believe they are simple) or Discussion (if you wish to analyze them).

The Discussion section must include a clear description of your conclusions. You could also conclude that the error bars are too big for you to conclude anything or that the procedure didn't work. But it is important that you take a stand-no waffling. The Discussion section should return to the questions, goals, and issues raised in your Introduction. Hence, the Introduction and Discussion are the bookends for the Results.

Conclusion

This is the summarization part or generalization part of the experiment and should be related with an objective.

References

Cite any references you used to do this in lab or in your discussion (other than what was presented in the lab manual).

Experiment 1: determination of refractive index and molar refraction

Objective: To determine the refractive index, specific refraction, and molar refraction of the provided liquid samples.

Theory

The refractive index, sometimes referred to as the index of refraction (nD), is a physicochemical property of substances (optical medium). We may learn about the behavior of light by examining the "refractive index." It is most likely that Thomas Young coined the term "index of refraction" in 1807 and used it initially. Because each material has a higher refractive index than the others, light travels through them at a slower speed. The reason for this could be the way light is affected by the interactions between molecules that make up the substrate. Furthermore, in the majority of substrates, rising temperatures cause the refractive index to drop. An increase in temperature reduces the amount of molecular interaction. Through the use of refractometers, the refractive index of various substrates is measured.

Although light moves through the vacuum at the same speed, it moves more slowly through other media because the atoms in that material are always absorbing and reemitting light. It is the ratio of the speed of light in a vacuum to the speed of light in that substance:

$$\text{Refractive index of a substance}(n) = \frac{\text{speed of light in vacuum}}{\text{speed of light in substance}} \tag{1}$$

Refraction is the process by which light changes the direction of travel along with its speed as it passes across a border between two media (Figure 1). (If light is moving perpendicular to the barrier, it enters the new medium without changing its direction.)

The relationship between the refractive indices of the two media (n_1 and n_2), the angles of incidence (θ_1) and refraction (θ_2), and the speed of light in the two media (v_1 and v_2) is given as follows:

$$\frac{v_1}{v_2} = \frac{\sin \theta_1}{\sin \theta_2} = \frac{n_2}{n_1} \tag{2}$$

n_1 and n_2 are the corresponding refractive indices, θ_1 and θ_2 are the angles of incidence and refraction, respectively, and v_1 and v_2 are the propagation velocities in the less dense media M_1 and M_2, respectively.

Therefore, measuring the speed of light inside a sample is not necessary to determine its index of refraction. However, through measuring the angle of refraction and understanding the index of refraction of the layer in contact with the sample, the refractive index of the sample can be precisely measured. Refractometers are devices used to measure the refractive index of different substrates. Another tool used for de-

https://doi.org/10.1515/9783111702261-002

termining the refractive indices of gases, is the Rayleigh refractometer. At a known wavelength of 589.6 nm, a sodium lamp can serve as the light source; however, many instruments require correction.

Refractometers can indeed be classified into four main categories based on their design and application:

1. Handheld refractometers: These are small, portable instruments that are frequently employed in brewing and agriculture. They are often easy to use for rapid measurements and feature a straightforward optical construction.
2. Abbe refractometers: Often used in labs, these devices offer excellent precision when determining the refractive index of both liquids and solids. They need a prism and a light source and are more complicated.
3. Digital refractometers: To determine the refractive index, these contemporary instruments use electronic sensors. They are easy to use because they provide fast readings and frequently have features such as temperature compensation.
4. Laboratory refractometers: Made for more specific laboratory uses, these refractometers have the ability to measure small samples with great precision and may be equipped with sophisticated capabilities such as data logging and automated temperature correction.

Each type has unique applications, benefits, and drawbacks according to the sample type and the level of precision needed.

The refractive index of liquids and solids can be measured with sophisticated devices called digital refractometers. Here are a few salient characteristics and advantages:

Features:

Electronic sensors: These devices provide accurate readings by using sensors to identify light refraction.

Digital display: Results are easy to read and understand because they are displayed on an LCD screen.

Temperature compensation: To provide precise readings, several models automatically adapt to temperature changes.

Calibration: For ease of setup and upkeep, digital refractometers frequently come with integrated calibration functions.

Compact design: The majority are easy to use and portable, making them appropriate for both laboratory and fieldwork applications.

Data logging: To monitor changes over time, certain models have the ability to retain repeated readings.

Applications:

Food and beverage: Used to gauge the amount of sugar in food preparation, winemaking, and brewing.

Pharmaceuticals: Assists in formulation development and quality assurance.

Chemical industry: Aids in quality assurance by measuring solution concentration.

Agriculture: Helps determine the concentration of plant sap for agricultural management.

Advantages:

Speed: Reduces the analysis time by providing fast readings.

Accuracy: Provides greater precision than conventional techniques.

Usability: Usually requires little training, so a wide range of people can utilize them.

Handheld refractometers

They are portable instruments made to measure a liquid's refractive index quickly and easily. The following are some important factors:

Features:

Portability: They are small and light, which make them perfect for fieldwork and measurements while on the go.

Easy operation: They usually have an eyepiece or a digital readout for convenient use and a simple design.

Scale types: A large number of portable refractometers come with particular scales for various uses such as sugar concentration scales for Brix or salinity scales for marine applications.

Optical design: Use lenses and prisms to bend light such that a scale may be used to visually assess the refractive index.

Durability: Frequently made to resist severe circumstances, they are appropriate for outdoor settings.

Applications:

Agriculture: Determines the concentration of plant sap or the amount of sugar in fruits.

Brewing and winemaking: Assists in tracking fermentation and figuring out how much sugar is in wort or must.

Aquaculture: Assesses water salinity to guarantee ideal circumstances for aquatic life.

Food production: Measuring sugar and other soluble solids in food products helps with quality control.

Advantages:
Instant results: Offers fast readings, which is crucial in settings when time is of the essence.

User-friendly: A broad spectrum of consumers can utilize them because they require little training to operate.

Economical: Usually less expensive than digital refractometers, making them ideal for hobbyists or small enterprises.

Disadvantages:
Accuracy: Although they work well for everyday applications, they might not be as accurate as equipment made for laboratories.

Manual calibration: May require manual calibration and adjustment for accurate readings.

Laboratory refractometers
They are specialist devices made to test liquids' and solids' refractive indices with great precision in controlled settings. The following are the main attributes and uses:

Features:
High precision: Designed for precise measurements, frequently with a four-decimal-place precision.

Advanced optics: To provide accurate and dependable readings, use premium prisms and light sources.

Temperature control: To guarantee accuracy in a range of circumstances, several models have integrated temperature correction mechanisms or temperature control features.

User interface: Usually furnished with digital screens, calibration buttons, and occasionally touchscreen controls for simpler operation.

Sample versatility: Capable of measuring a variety of samples, including tiny amounts of solids and liquids.

Uses:
Chemical analysis is the process of determining the refractive index of different chemical substances for use in research and quality control.

Pharmaceuticals: By guaranteeing appropriate constituent concentrations, they aid in formulation development and quality assurance.

Food and beverage: Determines how much sugar and other solids are present in goods such as wines, syrups, and juices.

Material science: Aids in describing optical materials and determining their composition and purity.

Advantages:
Accuracy and reliability: Provides excellent accuracy and repeatability, which are necessary for work in laboratories.

Data logging: A large number of models provide the ability to save measurements for reporting, analysis, or monitoring changes over time.

Comprehensive calibration: Changes can be made in accordance with particular standards or circumstances, thanks to advanced calibration options.

Limitations:
Cost: Typically more costly than computerized or pocket refractometers, which limits their use in small-scale settings.

Complexity: Compared to simpler models, this one could need more training to function properly and evaluate findings.

Because they provide accurate measurements that are vital for a variety of applications, laboratory refractometers are vital instruments in scientific research and quality control.

Abbe's refractometer
The most useful and widely used refractometer, the Abbe instrument, has an optical system that is schematically depicted in Figure 1.

Features:
Optical design: To measure the refractive index precisely, use a high-quality prism and a light source, usually a sodium lamp.

Scale readout: To improve readability, measurements are shown on a calibrated scale, frequently with a magnifying lens.

Temperature control: To ensure precise readings, several versions use an integrated thermostat or water jacket to keep the temperature steady.

Sample holder: Contains a sample well for convenient liquid sample placement, and certain versions may measure solid samples with the use of particular accessories.

Calibration: Enables either automatic or manual calibration to guarantee measurement precision.

Applications:
Chemical and material science: This field helps with characterization and quality control by calculating the refractive index of different chemicals and materials.

Pharmaceuticals: By calculating the refractive index of active substances and formulations, it helps in formulation development and quality evaluation.

Food and beverage: Used in labs to measure the amount of sugar in juices, syrups, and other goods.

Petrochemicals: Provides vital information for quality control and aids in the analysis of oils and other petroleum products.

Advantages:
High accuracy: They are crucial for scientific research because of their reputation for producing measurements that are both incredibly exact and reproducible.

Versatility: Able to measure a variety of materials, including solids and liquids.

Durability: Made of sturdy materials to resist frequent use in a laboratory.

Limitation:
Complexity: Compared to handheld devices, it takes more knowledge and training to use efficiently.

Cost: Typically more costly, they are best suited for industrial and research labs.

A thin layer (~0.1 mm) of the sample is positioned between two prisms. Because the upper prism is securely positioned on a bearing, the side arm denoted by dotted lines permits rotation. This hinge allows the top and bottom prisms to be separated for cleaning and sample insertion. An infinite number of rays that enter the prism through light reflection essentially originate from the lower face of the prism and go through the sample at all angles. This surface is referred to as rough ground.

When the sample and the upper prism's smooth ground face come into contact, the radiation is refracted there, and this is followed by entering the fixed telescope. Separating the divergent critical angle rays of different hues into a single white beam that matches the course of the sodium D ray is accomplished by means of two Amici prisms that are rotatable with regard to one another.

A crosshair is attached to the telescope's eyepiece, and while taking a measurement, the prism angle is adjusted until the light–dark interface is almost in line with the crosshair. Then, the fixed scale – which typically graduates in nD units – is used to determine the prism's position. By moving water through the jackets that encircle the prism, thermosetting is done. Measuring the shift in the direction of collimated light as it moves from one medium to another is typically used to calculate a substance's refractive index.

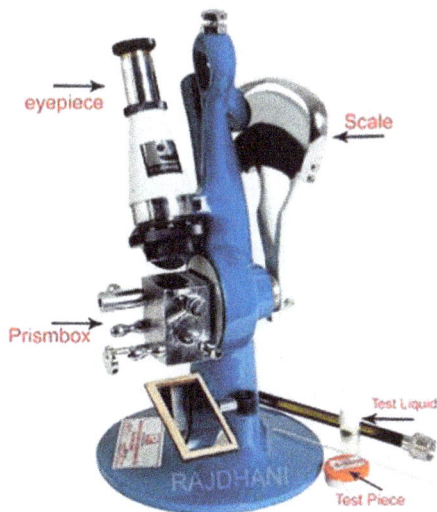

Figure 1: Abbe refractometer.

Since v_1 in eq. (1) becomes equal to c when M_1 is a vacuum, n_1 equals unity. Therefore,

$$n_2 = n_{vac} = \frac{\sin \theta_1}{\sin \theta_2} = \frac{c}{v_2} \tag{3}$$

The absolute refractive index of M_2 is represented by n_{vac}. By measuring the two angles, θ_1 and θ_2, one may thus determine n_{vac}. To find the organic liquid's refractive index, utilize Abbe's refractometer provided. The device is calibrated using a certain monochromatic light source and water as the liquid. To highlight the distinction between the bright and dark areas, use the micrometer screw. The cross wire of the telescope should be precisely on the edge separating the bright and dark sections after adjusting the refractometer scale. After calibrating the device, repeat the procedure with several organic solutions.

Liquid sample in the Abbe's refractometer is positioned between an illuminating and a refracting prism in a thin layer (see Figure 1). The refracting prism is made of a high refractive index glass, and the refractometer is meant to be used with samples whose refractive index is lower than that of the refracting prisms. Its bottom surface is textured like a ground glass joint, and the illuminating prism projects light onto it. Any point on this surface can therefore be thought of as creating light beams that are traveling in all directions. A detector placed behind the refracting prism would therefore show a region of light on the left and a region of darkness on the right. Because the prism's refractive index and angle of incidence are fixed, samples with varying refractive indices will exhibit varying angles of refraction.

Equation (2) shows that this will cause the border between the light and dark zones to move. By appropriately calibrating the scale, you can utilize the borderline location to determine the refractive index of any sample. This is the basic function of a real Abbe's refractometer; however, the rear of the refracting prism has different optics and no detector. A refractometer could instead be designed using the light reflection from the sample–prism boundary. Industrial processes are frequently monitored continuously using refractometers of this type. Light speed and index of refraction change dramatically with wavelength in the majority of liquids and solids.

Dispersion is the variance that enables white light passing through a prism to be refracted into a rainbow. Typically, longer wavelengths are refracted less than shorter ones. Monochromatic light is therefore required for the most precise measurements. The sodium D line at 589 nm is the most often used light wavelength for refractometry.

If white light were used in the simple Abbe's refractometer optics shown in Figure 1, the bright and dark boundaries would be at different places for different light wavelengths. Precise work would not be possible due to the consequent "fuzziness" of the borderline. Still, many Abbe's refractometers can work well with white light by inserting a sequence of "compensating prisms" into the optical path after the refracting prism. These compensatory prisms are designed to be able to be adjusted in order to correct, or compensate for, the sample's dispersion. This way, they can replicate the refractive index that would be produced with monochromatic light, namely the sodium D line at 589 nm.

As already mentioned, because light is absorbed and reemitted by the atoms in the sample, light moves more slowly in a substance than it does in a vacuum. It is not unexpected that the speed of light in a liquid will typically increase as the temperature rises since a liquid's density often decreases with temperature. As a result, for a liquid, the index of refraction typically drops with increasing temperature. The index of refraction of many organic liquids drops by about 0.0005 for every degree Celsius as the temperature rises. In contrast, the difference in water is only roughly −0.0001/°C. In order to maintain a specific temperature, many refractometers come with a thermometer and a way to circulate water through the device.

At 20 or 25 °C, the majority of refractive index measurements published in the literature are calculated. There are numerous applications for the refractive index. Its main applications are in material identification, purity verification, and concentration calculations. It is usually used to find the concentration of the solute in an aqueous solution. The sugar content (Brix degree) of a sugar solution can be determined using the refractive index.

It can also be applied in the pharmaceutical business to determine medication concentration. Lens focusing power and prism dispersive power calculations are made with it. It is also used to estimate the thermophysical characteristics of petroleum mixes and hydrocarbons. The rows above demonstrate how research on substrates' refractive indices can be helpful in a variety of contexts (both industrial and academic).

Apparatus:
– Temperature controller
– Light source and samples
– Medicine droppers
– Small beaker

Instrument: Abbe's refractometer

Chemicals
– Cotton
– Distilled water
– Ethanol

Procedure
1. The laboratory supervisor performed the calibration with distilled water first.
2. A glass dropper was used to distil water, which was then placed on the refractometer's prism surface.
3. To ensure temperature stability, it was left to stand for a minimum of 3–5 min.
4. The refractive index choice was selected using the turn mode selector.
5. Then, the eyepiece is focused until the crosshair is visible.
6. To align the crosshair and shadow line, turn the adjustment control counter clockwise. Additionally, the dispersion correction wheel is turned until perhaps a color-free shadow line appears.
7. Then, the shadow line is changed to align with the crosshair.
8. The shadow line is then repositioned to align with the crosshair.
9. A press of the temperature and red buttons was made.
10. The mode selector is set to the BX-TC position in order to further calibrate.
11. There was a press of the "read" display button. A 000.0 or 0.1 LED reading is typical. However, no reading will be indicated by a negative sign and a decimal point. Should this continue, the adjustment control unit needs to be turned on until a typical reading appears on the screen.
12. Just a little tweaking is needed. After measuring 000.0, proceed to read through the reticle crosshair to the shadow line. Until consistently normal readings are obtained, the procedure is repeated.

Determining the refractive index of the substance:
1) Next, remove the protective lens tissue by opening the prism assembly cover.
2) The surface of the measuring prism is cleaned with alcohol and cloth before any samples are added. Then position the mode selector to the proper location.
3) To place the shadowline at the bottom of the field of view, turn the adjustment control counterclockwise.
4) The crosshair for an accurate reading is in the center of the shadowline.

5) To focus on the crosshair, the eyepiece is also adjusted.
6) The coarse adjustment knob moves the shadowline to the crosshair reticle.
7) Any red or green hue at the shadow line's edge has been removed by rotating the dispersion correction wheel.
8) Calculating the molar refraction was done once these requirements were satisfied.
9) The excess unknown liquid and certain surplus samples were disposed of and handled in accordance with laboratory protocols.
10) Using detergent and water, the equipment sets were completely cleaned.

Observation:
Room temperature = ------------degree Celsius

S. no.	Given liquid	Refractive index of liquid sample
1		
2		
3		

Particular refractive index $R = \frac{n_2 - 1}{n_2 + 1} * \frac{1}{d}$

  Molar refractive index, $M_m = R*M$ (molecular mass of liquid)

Experiment 2: determining the percentage composition of unknown mixture by refractive index

Objective: To find out the composition of a given binary combination (ethanol–water) by Abbe's refractometer.

Theory: The percentage composition of a mixture of miscible liquids can be determined by measuring the refractive indices of the binary mixtures of known compositions. The values of refractive indices are plotted against the percentage composition at a particular temperature. After determining the solution's refractive index, the graph can be used to calculate the composition of the unknown mixing solution. The sample is contained as a thin layer (~0.1 mm) between two prisms. The dotted line-marked side arm of the upper prism enables it to rotate, thanks to a bearing that is securely fixed on it. In order to allow for separation for cleaning and sample introduction, the lower and top prisms are hinged. All angles of the sample are penetrated by an infinite number of rays that originate from the lower prism face, which is rough ground when light is reflected into the prism. The sample and the smooth ground face of the upper prism interface is where the radiation is refracted. It then goes through the fixed telescope after this. A single white beam that follows the same path as the sodium D ray is created by collecting divergent critical angle rays of various hues using two Amici prisms that are rotatable with regard to one another. Crosshairs are attached to the telescope's eyepiece; to take a measurement, adjust the prism angle until the light–dark interface is almost in line with the crosshairs. Next, the prism's position is determined using the fixed scale.

The process of thermosetting involves moving water through the jackets that encircle the prism. Typically, the refractive index of a material is ascertained by observing the shift in the direction of collimated light during its transition from one medium to another.

Determining the percentage composition of an unknown mixture using refractive index can be accomplished through a process involving calibration and calculations based on the refractive indices of the pure components in the mixture. Here is how you can perform this analysis step by step:

Apparatus:
- beaker
- pipettes
- calculator

Instrument:
- Refractometer

https://doi.org/10.1515/9783111702261-003

Chemicals:
- pure component (known substance in the mixture)
- unknown mixture.

Working procedure

1. **Measure the refractive indices of pure components:**
   Measure the refractive indices of each pure component in the mixture using the refractometer. Record these values (n_1 for component 1, n_2 for component 2, etc.).
2. **Prepare the unknown mixture:**
   Measure the refractive index of the unknown mixture using the refractometer. Record this value (n_{mix}).
3. **Use the linear relationship:**
   The refractive index of a mixture can often be estimated using a linear mixing rule $n_{mix} = x_1 n_1 + x_2 n_2 + \cdots + x_n n_{nn-}$, where x_{i-} is the mole fraction (or weight fraction, depending on your approach) of each component, and n_i is the refractive index of each component.
4. **Assume the total composition:**
   For a binary mixture, you can express one component in terms of the other:
   $x_2 = 1 - x_1$
   Substitute this into the equation to simplify

   $$n_{mix} = x_1 n_1 + (1 - x_1) n_2$$

   - Rearranging gives $n_{mix} = x_1(n_1 - n_2) + n_2$
   - From this, solve x_1:
   - $x_1 = \frac{n_{mix} - n_2}{n_1 - n_2}$
   - Calculate x_2 as $x_2 = 1 - x_1$
5. **Calculate percentage composition:**
   Convert the mole fractions to percentages by multiplying by 100.

Procedure
1. Take 11 test tubes and number them. Take 10 cm^3 of pure water in tube number 1 and 10 cm^3 of ethanol in tube numbers 11.
2. Take 1, 2, 3, 4, 5, 6, 7, 8, and 9 cm^3 of water separately in tubes numbered from 2 to 10, respectively, with the help of graduated pipette.
3. Add 9, 8, 7, 6, 5, 4, 3, 2, and 1 cm^3 of ethanol, respectively, in tubes numbered from 2 to 10. Shake each tube carefully to ensure uniform mixing.
4. Determine the refractive indices of pure water, ethanol, as well as their mixtures.
5. Also determine the refractive index of the given condition of unknown composition.
6. Plot a graph showing the percentage composition of the mixing solution versus the refractive indices to ascertain the unknown composition.

Experiment 3: measurement of sodium and potassium using flame emission spectrometry

Objective: To determine the content of sodium and potassium by flame emission spectrometry.

Theory

In order to test the low amounts of potassium and sodium in a solution, Bowling Barnes, David Richardson, John Berry, and Robert Hood created an instrument in the 1980s. Flame photometer is the term given to this device. Measuring the intensity of light released when a metal is added to a flame is the basis of the flame photometer's operation. Information about the element is provided by its wavelength, and information about its concentration in the sample is provided by the color of the flame. Atomic absorption spectroscopy (AAS) includes flame photometry as one of its subfields. Another name for it is flame emission spectroscopy. In the discipline of analytical chemistry, it is now an essential tool. Certain metal ions, such as sodium, potassium, lithium, calcium, and cesium, among others, can be measured for concentration using a flame photometer.

The metal ions are employed as atoms in the flame photometer spectrum. This method is known as flame atomic emission spectrometry (FAES) by the Committee on Spectroscopic Nomenclature of the International Union of Pure and Applied Chemistry (IUPAC). FAES or simply "flame photometry" (Figure 2) is a relatively new technique in instrumental analysis. Its roots are in Bunsen's flame color experiments, which were used to qualitatively identify specific metallic elements. For the analytical determination of trace metal ions in solution, atomic emission is a quick, easy, and sensitive technique. The approach exhibits extremely little interference from other elements due to the distinctive and narrow (ca. 0.01 nm) emission lines from the gas-phase atoms in the flame plasma. When analyzing diluted aqueous solutions without significant interferences, around ±1–5% relative precision and accuracy are typical.

Limits of detection may be quite low. Typically, the detection limits of "good" elements range from around 1 ng/mL to 1 µg/mL. The technique works with a wide range of metallic elements, although it works best with Li, Na, K, Rb, Cs, Ca, Cu, Sr, and Ba, which are easily excited to higher energy levels at the comparatively mild temperatures of some flames. In a flame, metalloids and nonmetals primarily form polyatomic radicals and ions rather than isolated neutral atoms. Therefore, with very few exceptions and under extremely specific circumstances, nonmetallic elements are not appropriate for analysis by flame emission spectroscopy.

In contrast to gravimetry, flame photometry is a very empirical rather than an exact method of analysis. Put differently, the process needs to be carefully and frequently calibrated. A multitude of experimental conditions determine how much light escapes the flame and reaches the detector. Therefore, accurate and regular calibra-

https://doi.org/10.1515/9783111702261-004

tion is required for best outcomes. The low-temperature flame only reaches about 1,700 °C, whereas oxygen and acetylene reach 3,100 °C. Only the elements that are most readily excited create significant emission from this flame. A simple narrow-band pass interference filter can be used to isolate a wavelength. This filter is de-signed to transmit just the strong and characteristic sodium doublet lines around 589.0 and 589.6 nm.

Because it only has one direct-reading output and can detect one element at a time, the device is known as a "single-channel" photometer. The instrument has to be calibrated for a new element, and the filter needs to be replaced. The apparatus in-jects the sample into a mixing chamber with a PTFE spray-impact bead and multiple PTFE baffles to combine the fuel, oxidant, and sample droplets using a capillary aspi-rator. Due to this combination, only the tiniest droplets of the sample mist are able to enter the burner; the majority of the sample that is inhaled is wasted. The sample so-lution is consumed at a rate of 2–6 mL/min.

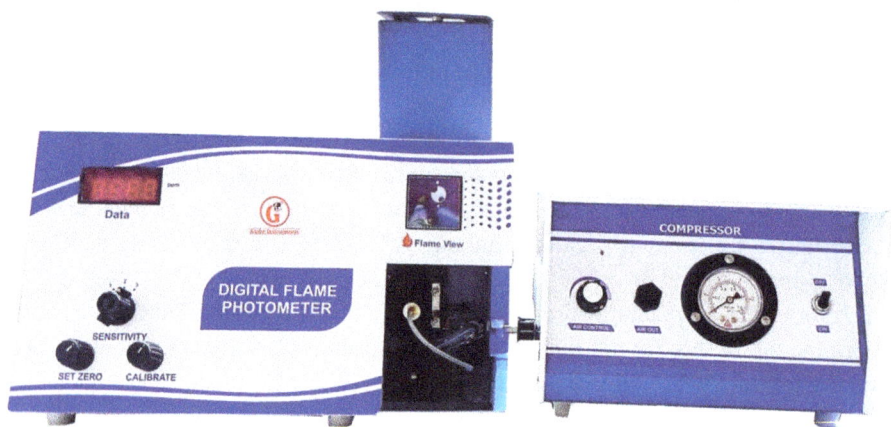

Figure 2: Microprocessor flame photometer graphical display.

A flame photometer does not need a light source because the light is produced by the sample's measured ingredient. The energy needed for the excitation is produced when acetylene or natural gas is burned in the presence of air or oxygen. The burners and aspirator on the instrument are the most sensitive parts. The gases are essential to the process of aspiration and aerosol production. The sample is drawn up by the air and directed into the aspirator, where larger droplets concentrate and may be re-moved. The monochromator chooses the ideal light wavelength to emit. One could uti-lize standard optical filters.

The light beam arrives at the detector. This photomultiplier generates an electri-cal signal in direct proportion to the light's intensity of emission. A solution of sodium and potassium samples sprayed into a flame will generate sample droplets, which will then be transformed by the flame's thermal energy into fine residue and, in the end, into the generation of neutral free sodium and potassium ions. The thermal energy of the flame transforms these free, neutral atoms back into excited state atoms. Atoms in the excited state emit a certain wavelength of radiation when they transition back to the ground state. A solution of sodium and potassium samples sprayed into a flame will generate sample droplets, which will then be transformed by the flame's thermal energy into fine residue and, in the end, into the generation of neutral free sodium and potassium ions. The thermal energy of the flame transforms these free, neutral atoms back into the excited state atoms. Atoms in the excited state release radiation at a particular wavelength when they transition back to the ground state.

For sodium (Na) = 589 nm and for potassium (K) = 466 nm.

The intensity of radiation released is dependent upon the concentration of an element in the sample, and the wavelength of radiation emitted characterizes a specific element in the sample.

The flame photometer's working principle

The complexes of the alkali and alkaline earth metals (group II) break up into individual atoms when they are introduced into the flame. Some of these atoms reach even higher levels of excitation. These atoms become unstable, nevertheless, in larger quantities. As a result, these atoms emit radiation when they return to their ground state. These radiations are usually found in the visible portion of the spectrum. Every alkali and alkaline earth element has a common wavelength.

Table 1: Emitted wavelength of metal with their corresponding colors.

Element	Emitted wavelength	Color of the flame
Sodium	589 nm	Yellow
Potassium	466 nm	Violet
Barium	554 nm	Lime green
Calcium	622 nm	Orange
Lithium	670 nm	Red

The number of atoms that return to the ground state is directly correlated with the intensity of the emission. Furthermore, the amount of light released correlates with the sample concentration.

Parts of flame photometer

The following fundamental parts make up a basic flame photometer:

The flame's source: The source of flame in the flame photometer is a burner. It is capable of being kept at a steady temperature. In flame photometry, one of the most important variables is the flame's temperature.

Nebulizer: A nebulizer is a device that introduces homogeneous solution into a flame at a steady pace.

The optical system: This is made up of a convex mirror and a convex lens. The light that the atoms emit is transmitted via the convex mirror. Convex mirrors aid in directing emissions toward the lens as well. The lens aids in concentrating light onto a slit or point.

Basic color filters: The mirror's reflections travel through the slit and onto the filters. The wavelength that has to be measured will be separated from extraneous emissions by filters.

Photodetector: This device measures the radiation intensity that the flame emits. Here, a light detector is used to transform the radiation that is released into an electrical signal.

Desolvation: In order to desolvate a sample, it must be dried in a solution. The flame dehydrates the metal particles in the solvent, which causes the solvent to evaporate.

Vaporization: The sample's metal particles likewise lose moisture. The solvent also evaporated as a result of this. The process of separating each atom in a chemical material is known as atomization. The flame turns the metal ions in the sample into metal atoms.

Excitation: The electrons' ability to absorb a specific quantity of energy is aided by the electrostatic force of attraction between the nucleus and electrons. Then, when excited, the atoms move to the higher energy state.

Emission: Atoms return to the ground state or low energy state in order to achieve stability because the higher energy state is unstable. This atomic hopping produces radiation with a distinct wavelength. The photodetector detects the radiation. Air, oxygen, and nitrous oxide are the three principal oxidants in flame photometers. The proportion of oxidant to fuel determines the flame's temperature.

Table 2: Temperature of some fuel–oxidant mixtures.

Fuel–oxidant mixture	Temperature (°C)
Natural gas–air	1,700
Propane–air	1,800
Hydrogen–air	2,000
Hydrogen–oxygen	2,650
Acetylene–air	2,300
Acetylene–oxygen	3,200
Acetylene–nitrous oxide	2,700
Cyanogen–oxygen	4,800

The processes occurring during flame photometer analysis are summarized below:

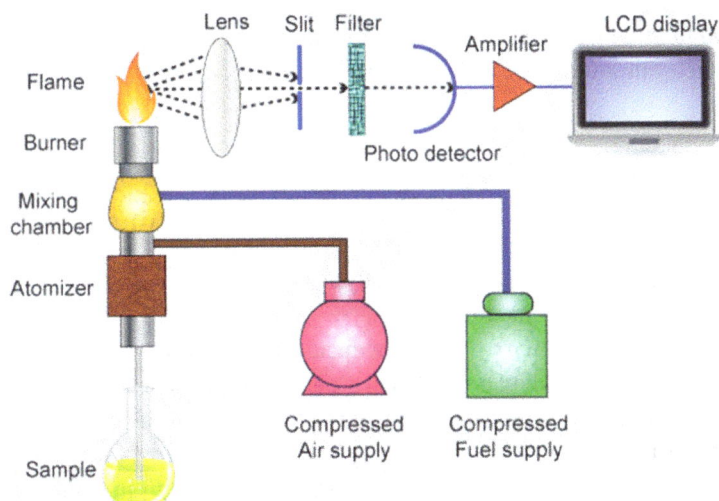

Scheme 1: Overview of flame photometry.

Scheibe–Lomakin equation

The Scheibe–Lomakin equation uses the following formula to describe the intensity of light emitted:

$$I = k \times c^n$$

where I is the intensity of emitted light, c is the concentration of the element, and k is the proportionality constant.

At the linear part of the calibration curve $n \sim 1$.

Thus, $I = k \times c$.

In other words, the intensity of the emitted light is directly related to the concentration of the sample.

Applications of flame photometer

Flame photometers are used for both qualitative and quantitative element analyses. The radiations that the flame photometer emits are unique to that metal. As a result, we can identify any particular element present in the sample by using a flame photometer. For soil health, a few group II elements must be present. After performing a flame test to identify the presence of certain alkali and alkaline earth metals in a soil sample, the soil can be fertilized with a particular type of fertilizer. The human body uses the concentrations of Na^+ and K^+ ions for a variety of metabolic processes. A blood serum sample can be diluted and aspirated into the flame to assess their concentrations. The amounts of different metals and elements can also be measured using flame photometry in soft drinks, fruit juices, and alcoholic beverages.

Advantages of flame photometer

–   The analysis procedure is relatively low cost and straightforward.
–   It is a sensitive, fast, practical, and selective analysis.
–   Its nature is both quantitative and qualitative.
–   It is possible to determine even extremely low metal concentrations in the sample, ranging from parts per million (ppm) to parts per billion (ppb).
–   Any unexpected interfering substance found in the sample solution is compensated for using this procedure.
–   It is possible to estimate elements that are rarely examined using this method.

Disadvantages of flame photometer

–   It is impossible to measure the metal ion's precise concentration in the solution.
–   It is unable to identify and detect inert gases directly.
–   Although the overall metal concentration of the sample is measured using this technique, the molecular structure of the metal present in the sample is not disclosed. Samples must be liquid to be used.
–   In some situations, sample preparation also becomes time-consuming.
–   It is not possible to determine the exact location of every metal particle using flame photometry. This technique is not able to analyze a large number of metal atoms.
–   Because they do not radiate, elements such as carbon, hydrogen, and halides are invisible.

Apparatus:
- Volumetric flask, pipette, beaker, funnel, filter papers, and spatula
- Wash bottle
- volumetric flask
- Assorted
- Graduated transfer pipets
- Small plastic containers

Instrument: Flame photometer

Chemicals
- Distilled water
- NaCl
- KCl

Calibration of instrument:
1. Connect the unit mains through the main unit.
2. Activate the power switch to turn on the digital display.
3. Set the selectivity switch to the desired setting by turning the coarse and fine knobs clockwise.
4. Feed the distilled water to the atomizer and wait for at least 30 s.
5. Set the course and fine-tune until the readout of the digital display approaches 0.
6. Feed the highest standard of the serial dilution and adjust till it displays 100.
7. Repeat steps 4–6 until the repeatability is attached.

Procedure for sodium

Measure the emission intensity for each standard, the unknown(s), and the blank (deionized water) carefully according to the instrument's operating instructions.
1. Use deionized water for the "blank."
2. Weigh accurately 100 mg of NaCl and dissolve it in 100 mL of distilled water.
3. Take 10 mL of the solution and dilute it to 100 mL with distilled water.
4. Prepare a series of standard solutions of 10, 20, 30, 40, and 50 µg/mL concentrations by diluting carefully to the mark with deionized water and mix thoroughly.
5. Switch on the instrument and select the sodium filter.
6. A nonluminous flame and air pressure between 0.4 and 0.5 kg/mL should be achieved by setting the gas in the flame.
7. Using the knob, atomize the flame strength to 0% with distilled water.
8. Atomize the flame intensity to 100% using 50 µg/mL standard solutions (highest concentration).

9. Measure the percent flame intensity of all the standard solutions (10, 20, 30, 40, and 50 µg/mL concentrations).
10. Plot the solution concentration and percentage of flame intensity on the graph.
11. Take the unknown from your instructor and carefully dilute it with deionized water to the 100 mL mark and stir well.
12. Mark the percentage flame intensity of the unknown sample on the graph, and then extrapolate to find the corresponding concentration.

Procedure for potassium

1. Weigh accurately 100 mg of KCl and dissolve it in 100 mL of distilled water
2. Take 10 mL of the solution and dilute it to 100 mL with distilled water
3. Prepare a series of standard solutions of 10, 20, 30, 40, and 50 µg/mL concentrations by diluting carefully to the mark with deionized water and mix thoroughly.
4. Switch on the instrument and select the sodium filter.
5. A nonluminous flame and air pressure between 0.4 and 0.5 kg/mL should be achieved by setting the gas in the flame.
6. Using the knob, atomize the flame strength to 0% with distilled water.
7. Atomize the flame intensity to 100% using 50 µg/mL standard solutions (highest concentration).
8. Measure the percent flame intensity of all the standard solutions (10, 20, 30, 40, and 50 µg/mL concentrations).
9. Plot the solution concentration and percentage of flame intensity on the graph.
10. Take the unknown from your instructor and carefully dilute it with deionized water to the 100 mL mark and stir well.
11. Mark the percentage flame intensity of the unknown sample on the graph, and then extrapolate to find the corresponding concentration.

Dilutions:
1. 100 mg dissolved in 100 mL water = 1,000 µg/mL
2. 10 mL diluted to 100 mL water = 100 µg/mL
3. 1 mL of 100 µg/mL diluted to 100 mL = 10 µg/mL
4. 2 mL of 100 µg/mL diluted to 100 mL = 20 µg/mL
5. 3 mL of 100 µg/mL diluted to 100 mL = 30 µg/mL
6. 4 mL of 100 µg/mL diluted to 100 mL = 40 µg/mL
7. 5 mL of 100 µg/mL diluted to 100 mL = 50 µg/mL

Observation table

S. no.	Concentration (µg/mL)		% FI	
	For Na	For K	For Na	For K
	0	0		
	10	10		
	20	20		
	30	30		
	40	40		
	50	50		
	Unknown	Unknown		

Result: The concentration of sodium ion in the given sample of NaCl solution was found to be . . . . µg/mL by flame photometry.

9. **Instrument alert**. There are moments when the aspirator compartment does not empty correctly.

    The sign to look out for is when the reading starts to fluctuate while you are aspirating some solution. Then, you aspirate deionized water, but the Na signal is still drifting and considerable. You can see some of the yellowish emission of Na if you open the small window on the flame chimney. This indicates that the aspirator compartment has to be "flushed," or drained once again. The drain Tygon tubing that is connected to the aspirator-burner's bottom can usually be gently moved to achieve this by wriggling it slightly below where it is attached. Ask the TA for help if you are still having issues.

10. When all the experiment is finished, clean the work areas thoroughly, aspirate deionized water to clear out the aspirator/burner, and allow a TA know that the instrument is ready to be shut down.

11. Rinse with deionized water all of the glass and plasticware provided for the experiment. Reposition the item inside the drawer after draining the water.

Hazardous waste disposal

In this experiment, *no* hazardous materials are generated or used. Our only options are weak solutions of regular table salt. While the cold water is running, pour *all* of the solutions down the sink drain. Use deionized water to thoroughly rinse any glass or plastic utensils you used.

Data treatment

Plot the emission intensities versus the concentration of Na to create a calibration curve. By comparing the sample's concentration to its emission intensity from the calibration curve, one can ascertain the sodium concentration in the unknown sample. It can be preferable to average all three values for each solution and derive a single final value for the unknown, or to derive three independent values for the unknown, each utilizing its "own" calibration curve, and then average the three values. This depends on the instrument's drift as well as other variables.

To find out if one strategy works better than the other, try both. If the plot seems to be somewhat linear, or at least the part that shows your unknown, you can fit the data using a linear least-squares method using the Excel LINEST tool. This will also give you some good fit parameters. Provide the "best estimate" for the mean sodium concentration in parts per million (µg/mL) along with the corresponding standard deviation of the measurement.

Experiment 4: determination of calcium concentration using atomic absorption spectrometer

Objective: To determine the concentration of calcium using atomic absorption spectrometer.

Theory

AAS is a spectroanalytical method that quantifies the components of a mixture by using free atoms in their gaseous form to absorb optical radiation. With AAS, more than 70 unique components of the solution can be found. The AAS technology is used in many industries, including water, food and beverage, clinical, and pharmaceutical analysis. AAS is also used in mining operations, for example, to determine the amount of precious metals in rocks. AAS uses a light house's specific electromagnetic radiation wavelength to detect elements in liquid or solid samples. The four parts of the AAS structure are the sample introduction area, light source, monochromator or polychromator, and detector. This makes the AAS structure extremely distinctive.

Electrothermal and spectroscopic flame atomizers are the most common types of atomizers utilized today. Atomic absorption involves aspirating a solution containing the target metal into a flame at a temperature between 2,000 and 3,000 °C. At first, the metal's electrons – in this example, calcium – are in their lowest possible energy state. Next, a thin visible light beam is directed through the flame. The light's wavelength has been chosen to match that atom's electron transition. The metal absorbs some of this radiation, which raises the energy level of the electrons. Beer's law can be used to connect the amount of radiation absorbed to the calcium ion concentration in the solution.

The atomic absorption apparatus uses a hollow cathode lamp as its light source, with calcium serving as the cathode's material of choice being the same metal under analysis. The calcium atoms become stimulated when heated, and their electrons move to higher energy states. Observable radiation is released when the electrons return to their previous positions. The energy difference between the electron levels of the Ca atom and the energy of the photons that are released are equal. The wavelength of 423 nm, which is associated with a pronounced electron transition in the Ca atom, is the one we will be measuring.

The Ca atoms in the solutions supplied into the atomic absorption flame will selectively absorb the 423 nm radiation that the Ca atoms in the hollow cathode lamp emit. Dilutions are made to create solutions containing 1, 2, and 5 parts per milligram of calcium using the standard calcium solution prepared last week. The absorbance of each solution is measured once it is inhaled into the atomic absorption spectrophotometer's flame. A calibration curve showing absorbance against calcium ppm is plot-

https://doi.org/10.1515/9783111702261-005

ted using this data. After diluting the unidentified calcium solution from last week's lab by a factor of 100, it is inhaled into the flame to evaluate its absorbance.

The calibration curve is then used to determine the ppm of calcium in the diluted unknown solution. Next, the calcium concentration in the first unidentified solution is computed. Atomic absorption involves aspirating a solution containing the target metal into a flame at a temperature between 2,000 and 3,000 °C. At first, the metal's electrons – in this example, calcium – are in their lowest possible energy state. Next, a thin visible light beam is directed through the flame. The light's wavelength has been chosen to match that atom's electron transition. The metal absorbs some of this radiation, which raises the energy level of the electrons. Beer's law can be used to determine how much radiation is absorbed based on the concentration of calcium ions in the solution. The cathode of a hollow cathode lamp, which uses calcium as the example metal under analysis, serves as the light source for the atomic absorption equipment. The calcium atoms become stimulated and their electrons move to higher energy levels when heated. Visible radiation is released when the electron concentration drops back to normal. The energy difference of the electron levels in the Ca atom is reflected in the energy of the photons that are released.

This 423 nm radiation corresponds to a notable electron transition in the Ca atom, and it is this radiation that we will be measuring.

The calcium atoms in the solutions supplied into the atomic absorption flame will selectively absorb the 423 nm radiation that the cathode lamp's Ca atoms sent out.

Calcium solutions of 1, 2, and 5 parts per million are prepared by diluting the standard calcium solution that was made last week. Measurements of absorbance are made for each solution when it is inhaled into the atomic absorption spectrophotometer's flame.

A calibration curve showing absorbance against calcium ppm is plotted using this data. After diluting the unidentified calcium solution from last week's lab by a factor of 100, it is inhaled into the flame to evaluate its absorbance. The calibration curve is then used to determine the ppm of calcium in the diluted unknown solution. Next, the calcium concentration in the first unidentified solution is computed.

Apparatus
–   Automatic pipettor 500 µL
–   Pipette
–   Volumetric flask
–   Beakers

Instrument
–   Shimadzu 6300 AAS AA/AE spectrophotometer

Chemicals
- Calcium carbonate
- Hydrochloric acid(6 M)
- Deionized water

Procedure
Preparation of calcium stock solution:
1. Weigh out approximately 0.252–0.312 g of dry primary standard calcium carbonate ($CaCO_3$, FW = 100.087) accurately, to the nearest 0.001 g.
2. Use a few milliliters of deionized water to rinse into a 250 mL volumetric flask.
3. Dissolve in a small volume of 6 M HCl (a few milliliters), dilute with deionized water to the appropriate level, and thoroughly mix to a volume of 250.0 mL, which contains 1,248 ppm of $CaCO_3$ or 500.0 ppm of calcium ions.
4. To prepare three standard (5, 2, and 1 ppm) calcium solutions and one diluted unknown calcium solution, two 100 mL volumetric flasks are required.
5. Transfer 10.00 mL of the standard calcium solution (500.0 ppm) into a 100 mL volumetric flask, top it off with distilled water, and thoroughly mix.
6. Write "50.0 ppm calcium solution" on the label.

5 ppm calcium: Fill a 100 mL volumetric flask with 10.0 mL of the 50.0 ppm calcium solution, then dilute with distilled water to the appropriate level. After thoroughly mixing, transfer about 50 mL to a 100 mL beaker or flask that has been cleaned and dried, marking it as 5 ppm calcium. Empty the volumetric flask of the leftover solution.

2 ppm calcium: Use distilled water to thoroughly rinse the 100 mL volumetric flask. Now fill the volumetric flask with 4 mL of the 50.0 ppm calcium solution, dilute it to the appropriate amount, and thoroughly mix. 50 milliliters should be poured into a dry, clean container and marked "2 ppm calcium." Empty the volumetric flask of the leftover solution.

1 ppm calcium: Use distilled water to thoroughly rinse the 100 mL volumetric flask. A clean, dry container should hold 50 mL after 1 mL of the 50.0 ppm calcium solution has been pipetted, filled to the mark with distilled water, and thoroughly mixed. Put a label on this: calcium solution, 1 ppm. Get rid of any leftover solution in the flask.

Unknown solution: 1.00 mL of an unidentified calcium solution should be pipetted into a 100 mL volumetric flask after it has been thoroughly rinsed with distilled water. Add enough water to dilute to the desired amount. Dilute the solution and thoroughly mix it. Write "unknown" on this flask.

Atomic absorption analysis of the four solutions

Check that the atomic absorption spectrometer is in auto zero (AZ button) mode with the capillary tube submerged in distilled water before aspirating any sample into the flame:

1. Press the AZ button after submerging the aspiration capillary tube in distilled water. This removes all calcium from the flame and sets the instrument to zero absorbance.
2. Insert the capillary tube for aspiration into the sample. Press the READ button to view the absorbance on the display screen once the lamp energy measurement has stabilized.
3. For every standard and the unidentified solution, repeat steps 1 and 2 above. Take note of the absorbance after aspirating the three diluted standards into the flame.

Plot the absorbance of each of the three standards against the ppm calcium ion for the calibration curve, and then calculate the slope of the straight line. Since there should be no absorption of 423 nm radiation at zero calcium concentration, the straight line should approach the origin. In order to measure its absorbance, aspirate your diluted unknown solution into the flame. Using the calibration curve, ascertain the ppm concentration of calcium in the original, diluted solution.

Data sheet

Concentration of calcium in the standard solutions (ppm)	Measured absorbance
0.000	0.000
1.00	
2.00	
5.00	
Concentration of unknown	

Calculations and report

1) Plot absorbance versus concentration (ppm) of calcium for the 1.00, 2.00, and 5.00 calcium standards to create the calibration curve. Connect your graph to this document.

Make sure the data includes both zero absorbance and 0 ppm Ca. Plot the graph using Excel.

Experiment 5: determination of complex ion composition by Job's method of continuous variation

Objectives:
- To ascertain the formula of the complex produced between the iron(III) ion and the thiocyanate ion, SCN, using Job's method of continuous variations.
- Calculating the complex's equilibrium constant and estimating its molar absorptivity value.

Theory

Covalent connections between ligands, or tiny molecules or ions, and metal cations result in the formation of coordination complexes. Recall that sharing a pair of electrons by two atoms is known as a covalent bond. One lone pair of electrons on the ligand provides the shared pair of electrons between the ligand and the metal cation in a coordination complex. In contrast to ionic bonds, which disintegrate into water to produce ions, covalent bonds typically do not break down into water solutions. In a solution of water, coordination complexes can exist as neutral molecules or polyatomic ions. A coordination complex is formed when one or more ligands bind to the metal cation, depending on the characteristics of the ligand and metal cation. The coordination number of the metal is the number of donor atoms from ligands that are joined to the metal cation. One to more than six coordination numbers are possible. Numerous transition metal compounds typically have coordination numbers of 4 or 6. Reversible reaction: An unknown number, "x," of thiocyanate ions, SCN^-, combine with the iron(III) ion, Fe^{3+}, to create a coordination complex:

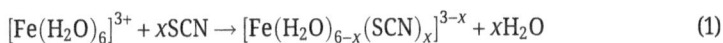

$$[Fe(H_2O)_6]^{3+} + xSCN \rightarrow [Fe(H_2O)_{6-x}(SCN)_x]^{3-x} + xH_2O \qquad (1)$$

The number of bonds that iron(III) may make with ligands is known as its coordination number; these are usually 1, 2, 3, 4, 5, or 6. Fe^{3+} has a coordination number of 6. Therefore, up to six monodentate ligands or three bidentate ligands can form bonds with iron(III).

For many years, the blood red complexes of Fe^{3+}/SCN^- have been utilized to measure trace levels of Fe(III) in aqueous solutions. $FeSCN^{2+}$ is the predominant species in extremely diluted solutions, whereas $Fe(SCN)_x^{3-x}$, higher complexes, are produced with increasing concentrations of SCN^-. At high SCN^- concentrations, some $Fe(SCN)_3$ is produced and can be partially removed into ether. In coordination chemistry, the hexathiocyanatoferrate(III) ion $[Fe(SCN)_6]^{3-}$ is well known. It has also long been known that the addition of inert electrolytes to a solution containing $FeSCN^{2+}$ or higher complexes results in a decrease in the intensity of color.

https://doi.org/10.1515/9783111702261-006

Every thiocyanate ion that is joined to an iron ion results in an overall charge of +3 − 1 for the coordination complex Fe $(SCN)_x^{3-x}$. Using Job's method of continuous variations, the experiment's initial step is to calculate the value of x. Reactant limitation is the foundation of Job's method of continuous variations. You will make several solutions in this experiment that contain varying volumes of liquids with known concentrations of SCN^- and Fe^{3+}. Based on the volume and concentration of the solution, one may determine how many moles of Fe^{3+} and SCN^- were added to create the final mixture. The number of moles of SCN^- and Fe^{3+} added to the solution is represented by the variables $nSCN$ and nFe. The mole fraction of SCN^- in the mixture can be ascertained as follows using the computed numbers of moles of Fe^{3+} and SCN^-:

$$xSCN = \frac{nSCN}{nFe + nSCN} \tag{2}$$

In the mixture that contains stoichiometric amounts of Fe^{3+} and SCN^-, it will be true, according to eq. (1), that

$$x(nFe) = nSCN \tag{3}$$

(Be careful not to confuse the coordination number, x, with the mole fraction symbolic notation χ.) Dividing both sides of eq. (3) by the quantity $nFe + nSCN$:

$$x\left(\frac{nFe}{nFe + nSCN}\right) = \frac{nSCN}{nFe + nSCN} \tag{4}$$

The right-hand side of eq. (4) is equal to the mole fraction of SCN, χSCN. The quantity in the parentheses on the left-hand side of eq. (4) is equal to the mole fraction of Fe, χFe. Making these replacements in eq. (4) one obtains:

$$x\chi Fe = \chi SCN \tag{5}$$

Since the sum of mole fractions in a mixture always adds up to one, the mole fraction of Fe is related to the mole fraction of SCN by the following equation:

$$\chi Fe = 1 - \chi SCN \tag{6}$$

Substituting eq. (6) into eq. (5), the following relationship is obtained:

$$x(1 - \chi SCN) = \chi SCN \tag{7}$$

Solving eq. (7) for x:

$$x = \frac{\chi SCN}{1 - \chi SCN} \tag{8}$$

Consequently, the mole fraction of SCN^- in the mixture can be used to compute the coordination number, x, for a combination having a stoichiometric amount of Fe^{3+} and SCN^-. The coordination complex of iron thiocyanate is red in color. As a result, it absorbs electromagnetic radiation in the visual range. As a result, the product's concentration can be determined by calculating the quantity of visible light it absorbs. An UV–Vis absorption spectrophotometer can be used to quantify experimentally how much visible light is absorbed by a substance.

The light intensity before it enters the sample, known as the incident light, or I_0, and the light intensity after it enters the sample, known as the transmitted intensity, or I_t, are measured by the device (see Figure 3). The transmitted intensity will be less than the initial intensity if a sample absorbs light at a specific wavelength. These two intensities are the basis for two calculations made by the spectrophotometer. The first is known as the transmittance percent, or %T, and it has the following definition:

$$\% \, T = \frac{I_t}{I_0} * 100 \tag{9}$$

A sample with 100%T does not absorb any light at that specific wavelength. A sample with 0%T completely absorbs all light falling on that specific wavelength. Generally speaking, a sample's %T falls between 0 and 100%T. The absorbance of a sample, A, is the other quantity that a spectrophotometer measures and is defined as follows:

$$A = - \log \left(\frac{I_t}{I_0} \right) = - \log \left(\frac{\%T}{100} \right) \tag{10}$$

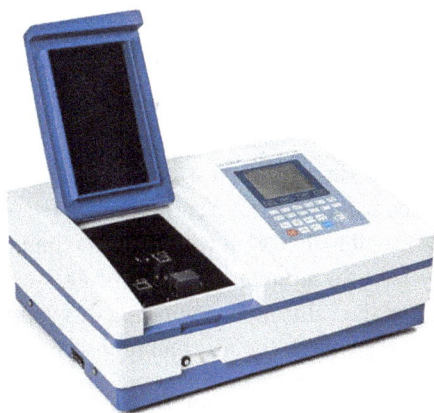

Figure 3: UV–visible spectrophotometer.

A sample's absorbance rises with the amount of light it absorbs. A sample that has no light absorbed is said to have an absorbance of 0.000. The amount that is often measured in a sample is the absorbance, or A. The concentration of the absorbing species is directly correlated with a solution's absorbance. The link between absorbance and concentration can be found in the following straightforward equation, which is also known as Beer–Lambert's law or more commonly as Beer's law:

$$A = \varepsilon l c \tag{11}$$

where ε is a proportionality constant known as the molar absorptivity or the extinction coefficient of the specific complex, and c is the concentration of absorbing species and l is the length of the sample that the light beam passes through. The inner diameter of the container (referred to as a cell) that contains the absorbing sample is the length of the sample, l, also known as the path length. This semester, the route length of our cells, is 1.00 cm in all of our applications. How well the sample absorbs light is determined by its molar absorptivity. The absorbing species and the light's wavelength determine its value.

In most cases, the absorbance of solutions containing known concentrations of absorbing species is measured experimentally, and the absorbance against concentration is plotted in a plot known as a Beer's law plot. Then, a line is fitted to the data points, and the line's slope equals the molar absorptivity. Lithosphere per mole per centimeter (L/mol cm) is the unit of measurement for molar absorptivity. By measuring the absorbance of a solution with a known concentration, one can experimentally determine the concentration if they are aware of the molar absorptivity of the absorbing species. Using Job's approach, you will first find the iron thiocyanate complex formula (the value of x) for this experiment.

Ten solutions with varying relative concentrations of Fe^{3+} and SCN^- will be measured for absorbance; these solutions' compositions are indicated by the equilibrium mixture compositions in Table 3. Using the specified volumes and concentrations of Fe^{3+} and SCN^- from Table 3, you will calculate the mole fraction of SCN^- for each solution. You will use a plotting tool (either Cricket Graph or Excel) to plot absorbance (the vertical axis) versus mole fraction of SCN^- (the horizontal axis) after measuring the absorbance of these 10 solutions. Let us say, for instance, that you get the set of data shown in Table 1. You should see a curve resembling the stoichiometric ratio in Figure 4.

To determine the mole fraction at which the maximum absorbance occurs, triangulate the plot as shown in Figure 4. The mole fraction of SCN^- at which the maximum absorbance occurs is the mole fraction in which the number of moles of Fe^{3+} and SCN^- are in the stoichiometric ratio. You can find the value of x by plugging this value of the mole fraction into eq. (8). For instance, in Figure 4, the maximum absorbance occurs at a mole fraction equal to 0.8:

$$X = \frac{0.80}{1 - 0.80} = \frac{0.80}{0.20} = 4 \tag{12}$$

Remember that x should be equal to a small whole number, so round off your answer to the nearest whole number.

Table 1: Sample data to be plotted in a Job's plot.

Mole fraction SCN⁻	0.0	0.1	0.2	0.3	0.4	0.5	0.6	0.7	0.8	0.9	1.0
Absorbance of complex	0.000	0.063	0.125	0.188	0.25	0.313	0.375	0.438	0.500	0.250	0.000

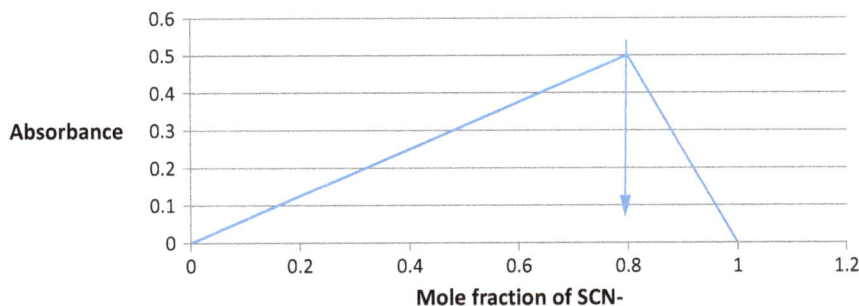

Figure 4: Typical Job's methods plot.

Finding the complex's molar absorptivity comes next, once the formula is understood. You will test the absorbance of five previously prepared solutions with known concentrations of iron thiocyanate complex to ascertain this. Plotting absorbance (the vertical axis) against concentration (the horizontal axis) will allow you to create a Beer's law plot. Fit the data to a straight line with a slope equal to the complex's molar absorptivity. Beer's law can be used to determine the equilibrium concentration of iron thiocyanate complex in one of your beakers using the computed molar absorptivity:

$$\left[Fe(SCN)_x^{3-x} \right] = \frac{\text{Absorbance}}{(\text{Molar absorptivity})(1.00 \text{ cm})} \tag{13}$$

Fe^{3+} and SCN^- equilibrium concentrations can be computed if the equilibrium concentration of the iron thiocyanate ion is known. First, the dilution formula is used to calculate the initial concentrations of SCN^- and Fe^{3+}:

$$\left[Fe^{3+} \right]_0 = \frac{(\text{Volume of } Fe^{3+} \text{ solution used})(0.0030 \text{ M})}{\text{Total volume of the solution}} \tag{14}$$

$$[SCN^-]_0 = \frac{(\text{Volume in liter of SCN solution used})(0.0030 \text{ M})}{\text{Total volume of the solution}} \tag{15}$$

One mole of Fe^{3+} and x moles of SCN^- are extracted from the solution for each liter of iron thiocyanate generated, in accordance with the chemical reaction shown in eq. (1). Consequently, the Fe^{3+} and SCN^- equilibrium concentrations will be equal to:

$$\left[Fe^{3+} \right]_{eq} = \left[Fe^{3+} \right]_0 - \left[Fe(SCN)_x^{3-x} \right] \tag{16}$$

$$[SCN^-]_{eq} = [SCN^-]_0 - x\left[Fe(SCN)_x^{3-x} \right] \tag{17}$$

The formation constant (or equilibrium constant) for the reaction can be calculated once the equilibrium concentrations are known:

$$K_f = \frac{\left[Fe(SCN)_x^{3-x}\right]}{\left[Fe^{3+}\right]_{eq}\left[SCN^-\right]_{eq}}$$ (18)

Apparatus
- Beaker
- Measuring cylinder
- Electronic balance

Instrument: Single beam UV–Vis spectrophotometer

Chemicals
- 0.1 M $AgNO_3$, solid NaF
- Fifteen 30/50 mL beaker, 0.25 M HNO_3 solution
- 0.20 M $Fe(NO_3)_3$ solution, 0.0030 M KSCN solution
- 500 mL volumetric flasks

Procedure
- **A) Table 2: Composition to Determine Molar Absorptivity**
- The volumes of reactants required to make the standard solutions are shown in the chart below. It is evident that the iron solution has a substantially higher concentration than the KSCN solution. This is to guarantee that the reaction uses up all of the KSCN. The volume and concentration of the KSCN (limiting reactant) utilized in each trial will be used to calculate the product's concentration.

Slope-Ratio Method. Because it may be used with systems where the complexes have significant dissociation constants and are therefore not well suited to the continuous variations or mole-ratio technique, this method is very useful. There are no lengthy, questionable extrapolations required in this method; instead, it is based on absorbance measurements of solutions when a substantial excess of one of the reactants inhibits the complex's dissociation. However, there are only a few situations in which the slope-ratio approach is appropriate. Beer's law must be adhered to, the system under study cannot give rise to more than one complex, and the complex must display distinctive absorption that sets it apart from its progenitors in order to be applicable.

The plot of Absorbance vs. Concentration exhibits a slope of ε if the experiment is performed with a significant "constant" excess of Mn^+.

SCN-has a far greater affinity (complexing) for Fe^{3+} than NO^{3-} does, and Fe^{3+} complexation with NO^{3-} is minimal. Therefore, by stabilizing the solution and increasing the interaction between SCN^- and Fe^{3+}, the acid is added. This is especially important at low SCN volumes, where a high acid concentration is required to make Fe^{3+} available for reaction with small volumes of SCN^-.

Solution number	Volume (mL) 0.0030 M KSCN	Volume (mL) 0.20 M Fe(NO$_3$)$_3$	Volume (mL) 0.25 M HNO$_3$	Total volume	Absorbance
1	1.00	25.00	74.00	100 mL	
2	2.00	25.00	73.00	100 mL	
3	3.00	25.00	72.00	100 mL	
4	4.00	25.00	71.00	100 mL	
5	5.00	25.00	70.00	100 mL	

Part I: determination of molar absorptivity

1. Adjust the Spectronics 20 spectrometer's wavelength to 447 nm. Once the spectrometer reads %T and the cell holder is empty and closed, turn the left-hand knob on the front until the reading drops to 0.0%.
2. Set the meter needle to read absorbance = 0.0 (= 100% transmittance) by turning the right front knob. Setting the "full scale" is the term for this stage.
3. Take a cell and give it two 0.25 M HNO$_3$ washes. Place the cell into the spectrometer, seal the cover, and add 0.25 M HNO$_3$ until it is roughly three quarters full. After changing the spectrometer's reading to absorbance, turn the front right-hand knob until the reading reaches 0.000 absorbance units.
4. The lab will have five solutions, each containing a particular volumetric of iron thiocyanate complex, available in 100 mL volumetrics. The quantities specified in Table 1 were used to prepare the solutions. Pour roughly 10 mL of each solution into each of the five small beakers, making sure to note which solution is in which beaker.

After cleaning the cell twice with small amounts of each solution, filling it approximately three-quarters full, inserting it into the spectrometer, and reading the absorbance are the methods used to measure and record the absorbance of each solution.

Table 3: Composition of equilibrium mixtures/formula of the complex.

Beaker Number	Volume (mL) 0.0030 M Fe^{3+}	Volume (mL) 0.0030 M SCN$^-$	Volume (mL) 0.25 M HNO$_3$	Total volume	Absorbance
1	0.00	16.00	4.00	20 mL	
2	1.00	15.00	4.00	20 mL	
3	3.00	13.00	4.00	20 mL	
4	5.00	11.00	4.00	20 mL	
5	7.00	9.00	4.00	20 mL	
6	9.00	7.00	4.00	20 mL	
7	11.00	5.00	4.00	20 mL	
8	13.00	3.00	4.00	20 mL	
9	15.00	1.00	4.00	20 mL	
10	16.00	0.00	4.00	20 mL	

You must dispose of all rinse solutions containing Fe^{3+} and SCN^- in a garbage container. Therefore, when you are finished with the experiment, pour all of your waste solution into a large beaker and pour the waste into the waste container. The small beakers should be cleaned, dried, and used in the following portion of the experiment.

Determination of the formula

Wash and label 11 small beakers, numbered 1 through 10. Verify the beakers' dryness. Three small 10 mL graduated pipets should be cleaned and labeled SCN^-, Fe^{3+}, and HNO_3. Use a small amount of the labeled solution to rinse each pipet, then use it to produce the following solutions in the numbered beakers, following Table 3's instructions. Use deionized water to rinse the cell. Try to shake the tube as much as you can. To get rid of any fingerprints from the side, use a Kim wipe to clean the exterior of the tube. To prevent fingerprints from getting in the light-transmitting areas of the cell, only handle the upper parts of it after wiping. Use deionized water to rinse the cell twice. Shake until dry, then rinse with a tiny amount of the solution from beaker 1 two or three times. Fill your cell approximately three quarters full with the beaker 1 solution after washing it two or three times. Record the absorbance of this solution by inserting the cell into the spectrometer. In beakers 2 through 10, repeat steps 5 and 6 for the remaining solutions.

Calculations

Part III: **determination of the formula**
Determine the starting number of moles of Fe^{3+} and SCN^- present in the solutions in beakers 1 through 10 from Part I based on the volumes and concentrations of the employed solutions. Utilizing the following formula, determine how many moles of Fe^{3+} and SCN_3 were added to each beaker:

$$nFe = (\text{Volume in liter of } Fe^{3+} \text{ solution used}) (0.0030 \text{ M}) \tag{19}$$

$$nSCN = (\text{Volume in liter of SCN solution used}) (0.0030 \text{ M}) \tag{20}$$

Calculate the mole fraction of SCN^- in beakers 1 through 10 using eq. (2).

Create a graph using a graphing application that shows the absorbance (vertical axis) against the mole fraction of SCN^- (horizontal axis). Get the graph printed. Triangulating the plot to determine the greatest absorbance will yield the mole fraction of SCN^- at that point. Plot the data and attach it to your sheet. Utilizing the mole fraction of SCN^- at maximum absorbance, find the complex's coordination number (the value of x). On your data sheet, write the complex's correct formula, including the complex's charge.

Part IV: **determination of molar absorptivity**

There is a significant excess of Fe^{3+} ions found in solutions 1 through 5 from Part I of the experiment. This basically indicates that the limiting reagent in this reaction is SCN^-. As a result, the iron thiocyanate complex concentration will be determined by dividing the initial SCN^- concentration by the coordination number x. Using the dilution formula ($V_1 M_1 = V_2 (total)M_2$), get the starting concentration of SCN^- in each solution:

$$[SCN^-]_0 = \frac{(\text{Volume in liter of 0.003 M KSCN solution added})(0.0030\ M)}{100.00\ mL} \tag{21}$$

Determine the amount of iron thiocyanate complex present in solutions 1 through 5 in the experiment's first portion:

$$\left[Fe(SCN)_x^{3-x}\right] = \frac{[SCN]0}{x}$$

Plot the absorbance (vertical axis) against the iron thiocyanate complex concentration (horizontal axis) on a square piece of paper, then fit the data points to a straight line. The iron thiocyanate complex's molar absorptivity is shown by the slope of this line.

Part V: **calculation of the equilibrium constant**

For Part V, consider the solution in beaker 5 (from Part II) only!

Determine Fe^{3+} and SCN^- starting concentrations using eqs. (14) and (15). Employing eq. (13), determine the equilibrium concentration of the iron thiocyanate complex in beaker 5 based on its absorbance, in accordance with Beer's law. Using eqs. (16) and (17), determine the equilibrium concentrations of Fe^{3+} and SCN^- from the balanced chemical reaction: Determine the value of K for this reaction by using eq. (18), which is the expression for the equilibrium constant.

Remark 1:

At a specific temperature, the value of K_c in eq. (18) remains constant. This indicates that solutions containing SCN^- and Fe^{3+} will react until the K_c value is satisfied by the concentrations of the remaining reactants and products. Regardless of the initial concentrations of SCN^- and Fe^{3+} employed, the same value of K_c will be produced. With the use of multiple solutions prepared in various ways, our goal in this experiment is to find the value of K_c for this reaction and demonstrate that K_c does, in fact, have the same value under all conditions. Because K_c is a handy, the reaction is a good one to examine.

Remark 2:

– Throughout the preparation of every mixture in this experiment, you will keep the H^+ ion concentration at 0.5 M. Although it does not directly contribute to the process, the hydrogen ion's presence is required to prevent the creation of brown-colored species like $FeOH^{2+}$, which could impede the analysis of $[FeSCN^{2+}]$.

- Add two to three drops of concentrated HNO_3 carefully to a $Fe(NO_3)_3$ solution. The acid will stabilize the solution and repress (suppress) the production of yellow hue.
- To gently acidify the KSCN solution, add a few drops of concentrated HNO_3. Note: This fix might not hold. Examine the solution before usage.
- The observed color shift provides proof of a chemical interaction between $Fe(NO_3)_3$ and KSCN. One can use a 0.1 M $AgNO_3$ solution to determine if SCN^- is in excess or not. If $AgNO_3$ is added in excess, the AgSCN complex will form and a colorless solution will result.

To determine whether or not Fe^{3+} is excess, utilize NaF solid. Excessive addition of NaF will result in a colorless solution (color fading) because of the $[FeF_6]^{3-}$ complex formation.

Experiment 6: determination of chlorides and sulfated by nepheloturbidometry

Objective: To determine chlorides and sulfates by nepheloturbidometry.

Theory
An analytical method called turbidimetry is used to quantify a liquid's turbidity, or cloudiness, which is usually brought on by the presence of suspended particles. It is possible to determine the concentration or size of these particles by quantifying the amount of light that is scattered by them as it passes through the liquid. Turbidimetry is the measuring of light transmission as a function of suspended particle concentration. The angle 180° is the measurement point for transmitted light intensity. Here is a breakdown of how turbidimetry works and its key components:

How it works:
1. **Light source:** A light beam passes through the sample, which is a liquid with particles in it.
2. **Light scattering:** Light is scattered by suspended particles as it travels through the liquid. The size and concentration of the particles have a direct relationship with the quantity of scattering.
3. **Detector:** The amount of light that enters the sample without being dispersed is measured by a detector positioned at a particular angle to the light source.
4. **Measurement of turbidity:** The sample's turbidity is determined by taking into account the drop in light intensity. The turbidity increases with the number of particles present.

Applications:
–   **Water quality monitoring:** Used to assess the clarity of drinking water or natural bodies of water. Higher turbidity levels can indicate pollution or the presence of microorganisms.
–   **Microbial growth:** In laboratories, turbidimetry is used to estimate bacterial or yeast growth by measuring the cloudiness of a liquid culture.
–   **Pharmaceuticals:** This helps in determining the stability of suspensions or colloids in pharmaceutical formulations.

Difference from nephelometry:
While **turbidimetry** measures the reduction in light intensity directly through the sample, **nephelometry** measures the scattered light at an angle, usually 90°. Both techniques are often used together for analyzing particulate matter in different applications.

Nephelometry is the measurement of light scattered as a function of suspended particle concentration (Figure 5). The relationship between concentration and scat-

https://doi.org/10.1515/9783111702261-007

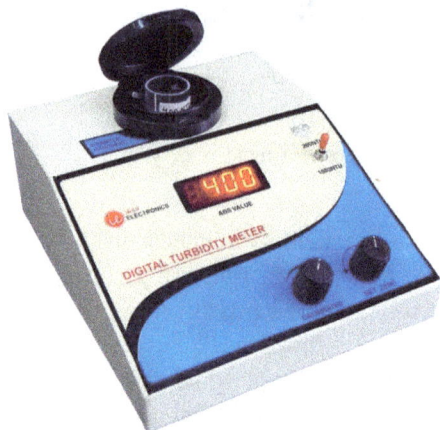

Figure 5: Turbidometer.

tered light intensity is proportional. Typically, the dispersed light intensity is expressed in terms of 90°. Measuring it at any convenient angle, such as 45°, 60°, or 135°, is also an option.

The intensity of transmitted light as a function of concentration, that is, when the concentration is more, it is less and when the concentration is less, it is more.

Apparatus
– Measuring cylinder
– Pipette
– Electronic balance
– Volumetric flask

Instrument
– Nepheloturbidometry

Chemicals
– Chloride standard solution
– 0.1 N silver nitrate
– Sulfate standard solution
– Dilute nitric acid
– Barium chloride solution, dilute acetic acid, and distilled water.

Procedure
Determination of chloride:
Preparation of standard opalescence:
1) Fill the Nessler cylinder with 10 mL of the chloride standard solution (25 ppm).
2) Increase the capacity to 50 mL with distilled water after adding 10 mL of diluted HNO_3.

3) To create the turbidity, add 1 mL of 0.1 N AgNO$_3$ solution and mix right away.
4) Let it stand in the shade for 5 min.

Preparation of sample opalescence:
1) Fill the Nessler cylinder with 10 mL of the sample solution using a pipette.
2) Fill the container to 50 mL with pure water after adding 10 mL of diluted HNO$_3$.
3) Quickly whisk in 1 mL of 0.1 N AgNO$_3$.
4) Let the finished mixture stand, shielded from light, for 5 min.

Measurement of nepheloturbidimetric unit (NTU)
1) Turn the device on
2) Give it 10 min to stabilize.
3) Add distilled water, then set the measurement to 0%.
4) Discard the water, take the standard opalescence, and adjust the reading to 100%.
5) Run sample opalescence and compare the % NTU with the standard.

Determination of sulfate:
Preparation of standard opalescence:
1) Fill the Nessler cylinder with 1.5 mL of the ethanolic standard sulfate solution using a pipette.
2) Add 1 mL of a 25% w/v barium chloride solution to this.
3) Stir and give it a minute to stand.
4) Add 0.15 mL of 5 M acetic acid and 15 mL of the standard sulfate solution (10 ppm) to this.
5) Use a glass rod to stir right away, then let it stand for five minutes.

Preparation of sample turbidity:
1) Fill the Nessler cylinder with 1.5 mL of the ethanolic standard sulfate solution using a pipette.
2) Add 1 mL of a 25% w/v barium chloride solution to this.
3) Combine and let stand for one minute.
4) Add 15 mL of the sample solution to this.
5) Include 0.15 mL of 5 M acetic acid.
6) Use distilled water to get the volume up to 50 mL.
7) Give it a little stir with a glass rod and let it stand for 5 min.

Measurement of nepheloturbidimetric unit (NTU):
1) Turn the device on
2) Give it 10 min to stabilize.
3) Add distilled water, then set the measurement to 0%.
4) Discard the water, take the standard opalescence, and adjust the reading to 100%.
5) Run the sample opalescence and compare the % NTU with the standard.

Observation tables

S. no.	For chloride	NTU range	Meter reading	Total NTu*meter reading
	Preparation			
1	Standard	1,000		
2	Standard			
S. no.	For chloride	NTU range	Meter reading	Total NTu*meter reading
	Preparation			
1	Standard	1,000		
2	Standard			

Experiment 7: determination of absorption maxima and effect of solvents on absorption maxima of organic compounds

Objective: To determine the absorption maxima and the effect of solvents on the absorption maxima of organic compounds.

Theory
The solvent dissolves the drug substance and exerts an intense influence on the quality and shape of the UV–Vis spectrum. Hence, the absorption spectrum of drug substance changes mostly as per the change of solvent that has been used to dissolve the drug substance. Here, the change in either wavelength (absorption maxima) or absorption intensity is monitored by changing the different solvents.

UV–Vis spectroscopy or UV/Vis, is the study of absorption or reflectance spectroscopy in portions of the electromagnetic spectrum that fall between the complete, nearby visible region and the ultraviolet. The field of absorption spectroscopy studies spectroscopic methods for measuring the amount of radiation absorbed by a substance as a function of wavelength or frequency. Photons, or energy, are absorbed by the sample from the radiating field. The absorption spectrum is the variation in absorption intensity as a function of frequency. As an analytical chemistry instrument, absorption spectroscopy is used to identify the presence of a certain item in a sample and, frequently, to quantify the amount of the material present. Analysis uses spectroscopies, especially those in the infrared (IR) and UV–Vis range, are very common. Molecular and atomic physics research, astronomical spectroscopy, and remote sensing all make use of absorption spectroscopy. To measure absorption spectra, a variety of experimental techniques are available.

Directing a produced radiation beam at a sample and measuring the radiation intensity that goes through it is the most typical setup. To compute the absorption, one can use the transmitted energy. Depending on the frequency range and the experiment's goals, the source, sample configuration, and detection method change considerably. Absorption maxima (λ_{max}), often known as lambda max, represents a material's strongest photon absorption wavelength at that point in the absorption spectrum. To put it simply, lambda max is the wavelength (Figure 6) at which a substance exhibits maximum absorption. Absorbances and absorption maxima can vary greatly among substances. To ensure that the detector receives a considerable amount of light energy, it is necessary to investigate substances that are highly absorbent in a diluted solution (absorbance value less than 1). This calls for the use of solvents that are entirely clear and do not absorb light. The solvents that are most frequently utilized are hexane, cyclohexane, water, and ethanol. In general, one should stay away from solvents with heavy atoms (such as S, Br, or I) or double or triple bonds.

https://doi.org/10.1515/9783111702261-008

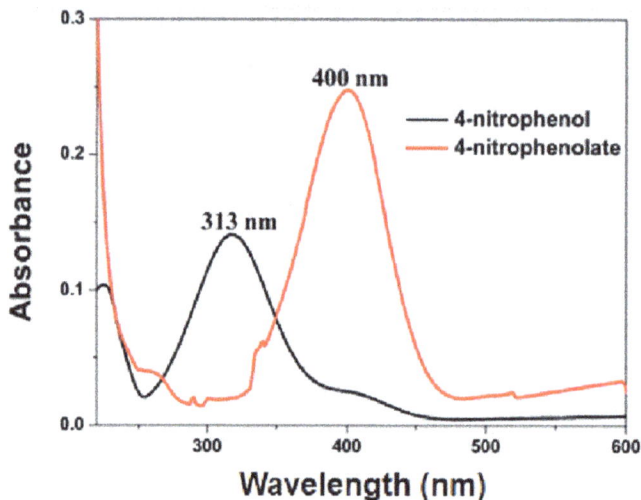

Figure 6: Typical example of an unknown sample depicting absorption maxima.

The Beer–Lambert

According to the Beer–Lambert equation, a solution's absorbance is directly correlated with the path length and the concentration of the absorbing species in the solution. Thus, UV/Vis spectroscopy can be used to find the absorber concentration in a solution for a specific path length.

Ultraviolet–visible spectrophotometer

A UV/Vis spectrophotometer is the tool used in UV–Vis spectroscopy. It calculates the light's intensity after it travels through a sample and contrasts it with the light's intensity prior to the sample.

Selection of solvents

Each solvent should have a wavelength at which UV–Vis absorbance cuts off. The wavelength below which all light is absorbed by the solvent is known as the solvent cutoff. Therefore, the absorbance cutoff of the solvent should be carefully considered by the student. Another solvent needs to be used if the solvent is exhibiting cutoff close to the material under investigation's absorption maxima.

Table 3: UV absorbance cutoff of some solvents.

Solvent	UV absorbance cutoff (nm)
Water	180
Ethanol	205
Toluene	285
Dimethyl formamide	267
Benzene	278
Acetone	329

Applications:

1. Analytical chemistry frequently uses UV/Vis spectroscopy to quantify a variety of analytes, including transition metal ions, highly conjugated chemical compounds, and biological macromolecules.
2. Because the electrons in the metal atoms can be stimulated from one electronic state to another, solutions containing transition metal ions have the ability to absorb visible light and become colored. The existence of additional species, such as specific anions or ligands, has a significant impact on the color of metal ion solutions. For example, adding ammonia to a diluted solution of copper sulfate increases the wavelength of maximum absorption (λ_{max}) and intensifies the color to a very light blue.
3. Organic compound, especially those with a high degree of conjugation, also absorbs light in the UV or visible regions of the electromagnetic spectrum.
4. UV/Vis can be applied to determine the kinetics or rate constant of a chemical reaction.
5. Spectroscopic analysis is commonly carried out in solutions but solids and gases may also be studied.

Apparatus

– Glass beakers
– Measuring flasks
– Whatman filter paper
– Measuring cylinder

Instrument

– UV–Vis spectrophotometer

Chemicals:

- Paracetamol
- Distilled water
- Ethanol, 0.1 N NaOH, and 0.1 N HCl

Procedure

1. Weigh precisely 10 mg of paracetamol, then dissolve it in enough solvent (2/3 volume), in 100 mL of measuring flask, shake well to dissolve completely, and makeup the volume up to the mark to prepare 100 ppm of stock solution.
2. To prepare a 10 ppm of solution, pipette out 1 mL of the stock solution, add it to a 10 mL measuring flask, and top it off with fresh solvent.
3. Scan the solution in UV–Vis spectrophotometer to obtain the absorption maxima.

Observation table

Solvent	λ_{max}
Distilled water	
Ethanol	
0.1 N NaOH	
0.1 N HCl	

Experiment 8: determination of chromium and manganese in a mixture

Objectives

– To calculate the concentrations of chromium and manganese in a mixture from the absorption measurements of the mixture at two different wavelengths, apply and adapt the method of simultaneous determination for other similar combinations of ions. Performing absorbance measurements for solutions on a spectrophotometer and drawing its UV–Vis spectrum, determine the wavelength of maximum absorption in the spectrum and compute the corresponding molar absorption coefficient.

Theory

When alloy steel containing manganese and chromium dissolves, the elements' ions – Cr^{3+} and Mn^{2+}, respectively – come out. Using potassium persulfate and potassium periodate, respectively, these are oxidized in the determination to produce dichromate and permanganate ions. The maximum absorption (λ_{max}) of the orange-red dichromate is measured at 440 nm, whereas the maximum λ of the pink permanganate is measured at 545 nm. Permanganate does, however, also absorb somewhat at 440 nm.

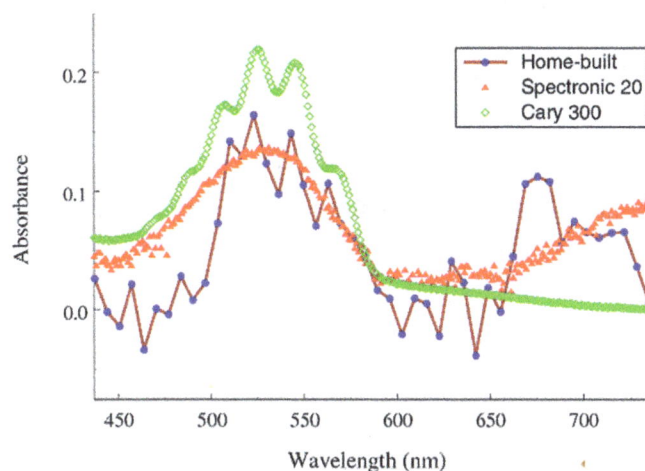

Figure 7: Schematic diagrams of the visible spectra of an 8.3×10^{-5} M potassium permanganate solution obtained with the Cary 300, the home-built spectrophotometer, and the Spectronic 20 spectrophotometer.

https://doi.org/10.1515/9783111702261-009

Simply expressed, both species absorb at the previously listed wavelengths of maximum absorption. Consequently, the solution containing a dichromate and permanganate ion mixture would absorb the total of the two species' absorbances at these wavelengths. When tested alone and in combination with 0.5 M sulfuric acid, it can be demonstrated that the absorbances of these two ions follow Beer–Lambert's law. With these parameters, and assuming a unit path length, we may write the following expression:

$$A_{440} = \varepsilon_{Cr,440}\left[Cr_2O_7{}^{2-}\right] + \varepsilon_{Mn,440}\left[MnO_4{}^-\right] \tag{3.1}$$

$$A_{545} = \varepsilon_{Cr,545}\left[Cr_2O_7{}^{2-}\right] + \varepsilon_{Mn,545}\left[MnO_4{}^-\right] \tag{3.2}$$

The general expressions for the simultaneous equations are

$$A_{\lambda1} = C_1(\mathcal{E}_1)\,\lambda_1 + C_2(\mathcal{E}_2)\,\lambda_2$$

$$A_{\lambda2} = C_1(\mathcal{E}_1)\,\lambda_1 + C_2(\mathcal{E}_2)\,\lambda_2$$

The definitions of the terminologies used here are clear. For instance, $\varepsilon_{Cr,545}$ denotes the molar absorption coefficient of chromium as dichromate ion at 545 nm; A_{440} denotes the mixture's absorbance at 440 nm; $[Cr_2{}^{2-}O^{7-}]$ denotes the concentration of seven dichromate ions, and so forth. To identify both species without separating them, we must solve these simultaneous equations. The following formulas for the concentrations of dichromate and permanganate ions are obtained by solving eqs. (3.1) and (3.2). If we are aware of the molar absorption coefficients for each of the two ions at these wavelengths, we can therefore determine the amounts of the two ions from the absorbance readings at 440 and 545 nm.

Measurements of the absorbance of pure solutions of the two compounds at the relevant wavelengths yield the values of the molar absorption coefficients. Since Fe^{3+}, Ni^{2+}, Co^{2+}, and V^{2+} absorb in these areas as well, the analyte solution should not contain any of these.

Apparatus
- Matched cuvette
- Volumetric flasks
- Volumetric flasks
- Graduated pipette
- Beakers
- Weighing bottle
- Burettes

Instrument
- Spectrophotometer

Chemicals
- Potassium dichromate
- Potassium permanganate
- Sulfuric acid

Solutions provided

i)   **1 M sulfuric acid (reference blank):** This is made by filling a 1-L flask with approximately 500 cm^3 of distilled water, adding 53.3 mL of analytical-grade concentrated sulfuric acid (while continuously whirling the flask), allowing it to cool, and then adding more distilled water to make up the volume.

ii) **0.002 M potassium dichromate in 1 M sulfuric acid**: It is made by quantitatively transferring 0.1471 g of analytical-grade potassium dichromate to a 250 mL volumetric flask, adding enough 1 M sulfuric acid to dissolve it, and then precisely weighing the result. About 1 M sulfuric acid is added to the volume to bring it up to the required level after the chemical dissolves.

iii) **Oxalic acid (0.005 M):** This is made by quantitatively transferring 0.063 g of oxalic acid to a 100 cm^3 volumetric flask, adding enough water to dissolve it, and then measuring and recording the results. Once dissolved, distilled water is added to bring the volume up to par.

iv) **0.002 M of potassium permanganate in 1 M sulfuric acid:** This is prepared as follows:

   a. Fill a 100 mL beaker with 50 cm^3 of distilled water, then weigh out about 1.8 g of potassium permanganate A.R.

   b. Bring the solution to a boil and simmer it for approximately 15 min.

   c. After the solution has cooled, strain it through a glass wool stopper in the funnel.

   d. Utilizing 1 M sulfuric acid, dilute 10 mL of the filtrate in a 100 mL beaker to make it 100 mL.

   e. Titrating the solution in an acidic medium with a standard oxalic acid solution will standardize it.

   f. Determine the volume of solution needed to make 250 mL of 0.002 mm solution.

   g. Transfer the necessary volume of the potassium permanganate standard solution into a 250 mL volumetric flask. Make up the volume with 1 M sulfuric acid.

v)   **Sample solution:** This is made by properly combining 10 mL of the 0.002 M potassium dichromate and potassium permanganate solutions, then adding 10 mL of 1 M sulfuric acid to the mixture.

Procedure

There are three main steps to the experiment's protocol. These are listed below.

A) Calculating the molar absorption coefficients of permanganate and dichromate ions.

B) Determining the additivity of absorbance values of permanganate and dichromate ions.

C) Finding the concentrations of permanganate and dichromate ions in the mixture.

Follow the instructions given below in sequential order to accomplish these tasks.

a) Determination of molar absorption coefficients for dichromate and permanganate ions

1. Transfer 20 and 40 cm³, respectively, of the potassium dichromate stock solution (0.002 M) into each of two 100 cm³ standard flasks. Fill the remaining volume with 1 M sulfuric acid. This would result in potassium dichromate solutions of 0.0004 M and 0.0008 M, respectively. Label the flasks appropriately.

2. Pipette 20 and 40 cm³ of the potassium permanganate stock solution (0.002 M) into each of two 100 cm³ standard flasks, and fill the remaining volume with 1 M sulfuric acid. This would result in potassium permanganate solutions of 0.0004 M and 0.0008 M, respectively. Label the flasks appropriately.

3. Calculate the absorbance values at 440 and 545 nm for the produced dilutions and the stock solutions (0.002 M) of potassium dichromate and potassium permanganate in steps 1 and 2. Make measurements in cuvettes of 1 cm path length and use 1 M sulfuric acid as the reference. Record the values in Observation Table 3.1.

4. Using the formulas provided under columns 4 and 6, calculate the molar absorption coefficients of potassium dichromate and potassium permanganate at 440 and 545 nm, respectively, and note the results in Observation Table 1.1.

b) Establishing the additivity of absorbance values of dichromate and permanganate ions at 440 and 545 nm

1. Fill two sterile burettes with 0.0004 M potassium permanganate and 0.0008 M potassium dichromate, respectively.

2. Using the information provided in columns 2, 3, and 4 of Observation Table 1.2, transfer the solutions of 0.0008 M potassium dichromate, 0.0004 M potassium permanganate, and 1 M sulfuric acid into seven beakers with a capacity of 100 cm³, labeled 1 through 7.

3. Record your findings in columns 5 and 7 of Observation Table 1 by measuring the absorbance values of the resultant solutions using 1 M sulfuric acid as reference.2. You have to utilize cuvettes with a 1 cm path length.

4. Using eqs. (3.1) and (3.2), respectively, determine the absorbance values for the mixtures of potassium dichromate and potassium permanganate at 440 and 545 nm, and note them in columns 6 and 8 of Observation Table 1.2.
5. To verify additivity, compare the computed and actual absorbance values. After Observation Table 1.2, note your observations in the space provided.

c) Determination of the concentrations of dichromate and permanganate ions in the mixture.

1. Use a cuvette with a path length of 1 cm to measure the absorbance values of the sample solution that has been provided.
2. Using eqs. (3.3) and (3.4), respectively, determine the mixture's potassium dichromate and potassium permanganate concentrations.

Observations and calculations

A. Determination of molar absorption coefficients for dichromate and permanganate ions.

Observation Table 1.1 Molar absorption coefficient measurements for $KMnO_4$ and $K_2Cr_2O_7$ at 440 and 545 nm.

S. no.	Concentration of $K_2Cr_2O_7$	Absorbance at 440 nm	Molar absorptivity at 440 nm	Absorbance at 545 nm	Molar absorptivity at 545 nm
1	0.002				
2	0.0004				
3	0.0008				
			Average =		Average =

Potassium permanganate

S. no.	Concentration of $K_2Cr_2O_7$	Absorbance at 440 nm	Molar absorptivity at 440 nm	Absorbance at 545 nm	Molar absorptivity at 545 nm
1	0.002				
2	0.0004				
3	0.0008				
			Average =		Average =

B. Determination of the absorbance values of dichromate and permanganate ions at 440 and 545 nm that are additive.

Observation Table 1.2 The potassium dichromate and potassium permanganate combinations' computed and observed absorbance values.

S. no.	Volume of 0.0008 M $K_2Cr_2O_7$ (mL)	Volume of 0.0004 M $KMnO_4$ (mL)	Volume of 1 M H_2SO_4 (mL)	Absorbance at 440 nm		Absorbance at 545 nm	
				Observed	Calculation	Observed	Calculated
	50	0	1				
	40	10	1				
	30	20	1				
	25	25	1				
	20	30	1				
	10	40	1				
	0	0	1				

C. Determination of the concentrations of dichromate and permanganate ions in the mixture.

Experiment 9: fluorometric determination of riboflavin in energy drinks

Objective: To measure the concentration of riboflavin in energy drinks

Theory

Riboflavin, also known as vitamin B_2, is a water-soluble vitamin that is found in dairy products, meat, fish, and some fruits and vegetables, particularly those with a deep green hue. Within a few days of dietary restriction, riboflavin, which is biochemically significant, shows signs of depletion. Athletes and other physically active individuals may need higher doses of B vitamins, particularly riboflavin, because activity puts strain on their metabolic pathways. These rationales clarify why riboflavin is commonly used in energy drinks. Many energy drinks, like Rockstar and Monster, offer serving sizes that are at least equal to the daily recommended amount of 1.3 mg.

Fluorescence

A photon, or light wavelength, is released when a material relaxes from an electrically excited state back to the ground state. We call this process luminescence. Though there are several ways to activate a material, in this experiment, we are interested in photoluminescence, or PL, as the form of excitation that arises from the absorption of a shorter wavelength photon. Luminescence occurs very infrequently. The majority of materials relax to their ground state via nonradiative pathways, which means they do not release any photons during this process.

Spin-allowed radiative transitions, which produce fluorescence in photoluminescent molecules, typically occur quickly (10^{-9}–10^{-8} s) between a singlet excited state (S1) and a singlet ground state (S0). Alternatively, luminescence may result from spin-forbidden radiative transitions from a triplet excited state (T1) to a singlet ground state (S0). The rate of this process, known as phosphorescence, is extremely slow – between 10^{-3} and 10^2 s for an average organic molecule. Nonradiative relaxation pathways virtually eliminate phosphorescence in solution at ambient temperature. Due to spin-forbidden singlet–triplet transitions, most molecules are singlet in their ground state.

Therefore, the only way to populate triplet excited states from singlet excited levels is by intersystem crossing (ISC). The emission fluorescence of a riboflavin sample is measured at 530 nm after it is excited at 447 nm.

Apparatus
- Volumetric flask
- Beaker
- Pipette

https://doi.org/10.1515/9783111702261-010

Instrument
- Fluorometer

Chemicals
- Quinine sulfate
- 0.1 N H_2SO_4
- Distilled water

Procedure

Preparation of standard solution:
1) Weigh the powdered medication, 100 mg of quinine sulfate, precisely.
2) To obtain the stock solution (1,000 µg/mL), dissolve in 100 mL of 0.1 N H_2SO_4.
3) Using 0.1 N H_2SO_4, dilute 10 mL of the stock solution to 100 mL (100 µg/mL).
4) Once more, add 1 mL of the solution above, and then dilute with 0.1 N H_2SO_4 (10 µg/mL) to make 100 mL.
5) Prepare concentrations of 0.5, 1, 1.5, 2, and 2.5 µg/mL from the solution above, and then use 0.1 N H_2SO_4 to dilute it to 10 mL.

Preparation of sample solution:
1) Using 0.1 N H_2SO_4, pipette out 1 mL of the provided sample solution to make the volume up to 10 mL.
2) Turn on the device and adjust the emission and excitation filters to operate between 360 and 460 nm in wavelength, respectively.
3) Use the highest concentration of the standard solution, 2.5 µg/mL, to set the fluorescent intensity (FI) to 100% and set the blank to 0.1 N H_2SO_4.
4) To prevent instrumental mistake, repeat the identical procedure two or more times.
5) Calculate the percentage of FI in various samples and standard solutions.
6) Plot the concentration versus FI graph and use the obtained FI to extrapolate the concentration of the unknown sample.

Experiment setup

Once Vernier Spectral Analysis is open, choose concentration versus fluorescence. A 405 nm excitation wavelength should be chosen. Keep in mind that the ideal excitation wavelength for riboflavin is 450 nm, as it is its maximal excitation wavelength. Still, the 405 nm LED in stock will function fine. Use a 450 nm LED in its place if one is available. Choose 526 nm as the wavelength of emission.

Zeroing the fluorimeter

Allow the device to be steady for a few minutes before starting to take measurements. After a few minutes, the device has to be calibrated. Go to the gear on the upper right corner of the window to accomplish this. Press "Calibrate" after inserting your blank – buffer or water – into the cuvette holder, choosing an integration period (100 ms is recommended), temporal averaging (3-to-5 is recommended), and verifying the excitation wavelength. There will be a pop-up window. After inserting your blank into the device, choose "Finish Calibration." Your spectrometer should be "zeroed" after it is finished.

This is shown by a red reading in the window's lower corner that is roughly 0.000. You will have to re-zero the fluorimeter if you discover later on that you need to adjust your settings (such as the integration time). You need to measure all of your standards and unknowns using the same acquisition parameters.

Acquiring fluorescence readings

It is easier to note the fluorescence intensity (Fl. Int.) for each sample in Table 2 rather than using the software to collect data. After inserting a fresh sample into the fluorimeter, give it ten to twenty seconds to stabilize. Try to estimate a center reading as the figures may vary somewhat. Repeat this process a minimum of three times for each sample.

Observation table

S. no.	Concentration (µg/mL)	%FI
	0	
	05	
	1	
	1.5	
	2	
	2.5	
	Unknown	

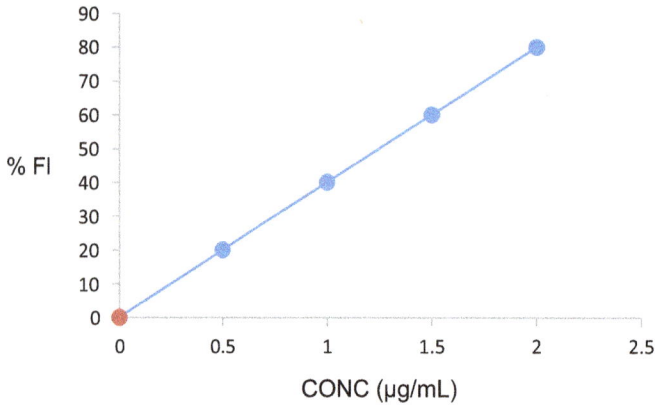

Concentration of unknown sample = ... µg/L

Data analysis

Plot the standard concentration against the average fluorescence intensity using Excel. Utilizing a linear regression to fit the data, note the coefficient of determination and the equation of the line. Determine the concentration of your unknowns and report the amount of riboflavin present in each unknown with the proper number of significant figures, just as you would with any external calibration method.

Experiment 10: Infrared spectroscopy of solids and solutions

Objective: To determine the molecular structure of compound in a given solid and solution using an IR spectrometer.

Theory

By analyzing how molecules absorb IR light, IR spectroscopy is a potent analytical tool used to identify and investigate chemical structures. The method's capacity to yield details about molecular vibrations, functional groups, and bonding environments makes it particularly helpful for characterizing both solids and liquids.

Principle of infrared spectroscopy

The basis of IR spectroscopy is the way in which matter interacts with this light. Certain frequencies of IR light, which correlate to the vibrational energy of their bonds, are absorbed by molecules. Molecular vibrations brought on by these absorptions result in the stretching or bending of chemical bonds. The "fingerprint" of the molecule is created by the IR spectrum, which shows the absorbed light as a function of wavelength or frequency:

- Wavenumber range: The typical range of operation for IR spectroscopy is 4,000–400 cm^{-3} (wavenumbers, cm^{-1}).
- Vibration types: Many molecular vibrations cause IR absorption, but the main ones are as follows:
- Stretching (change in bond length)
- Bending, or altering the binding angle

IR spectroscopy of solids

When dealing with solids, special techniques are used to record IR spectra. Solids present a challenge due to their opacity and scattering properties, which require tailored sample preparation.

1. Sample preparation techniques:

KBr pellet method: The solid sample is finely ground with potassium bromide (KBr), a transparent material in the IR region, and pressed into a pellet. This allows the IR beam to pass through the sample.

https://doi.org/10.1515/9783111702261-011

Attenuated total reflectance (ATR): A popular technique where the solid sample is placed on a crystal surface, and the IR beam reflects off it, penetrating the sample surface slightly to collect the spectrum.

Diffuse reflectance (DRIFTS): For powders or rough surfaces, IR light is scattered off the surface of the sample, and the diffuse reflection is measured.

Application in solid state:
Polymers: IR spectroscopy can be used to study the crystalline or amorphous nature of polymers.

Inorganic solids: It is useful for identifying materials like minerals, ceramics, and catalysts based on the unique absorption patterns of their chemical bonds.

IR spectroscopy of solutions

Since both the solvent and the solute may interact with IR light, solutions are simpler to investigate in IR spectroscopy. To prevent interference from solvent absorption, care must be exercised while choosing a solvent.

1. Sample preparation:
Liquid cells: For transmission measurements, solutions (such as CaF_2 or NaCl) are frequently put in liquid cells with IR-transparent windows. Usually, the path length is short (between 0.1 and 1 mm) to prevent total solvent absorption.

The solvents: Since they have little absorption in the IR spectrum, common IR-transparent solvents include carbon tetrachloride (CCl_4), carbon disulfide (CS_2), and deuterated solvents such as deuterated chloroform ($CDCl_3$).

2. Applications in solutions:
Organic molecules: The distinctive absorption bands of functional groups such as alcohols, ketones, and carboxylic acids make them clearly recognizable. Proteins and Biomolecules: By examining amide I and II bands in aqueous solutions, IR spectroscopy aids in the secondary structure analysis of proteins.

Key features in IR spectra:
Fingerprint region (1,500–500 cm^{-3}): This area serves as a unique identifier for every compound. Region of Functional Group (4,000–1,500 cm^{-1}): In this region, particular functional groups such as –OH, –NH, –CH, and –CO show distinctive peaks.

Peak analysis:
- O–H stretch: Wide peak centered between 3,200 and 3,500 cm^{-3} (caused by water or alcohol hydrogen bonding) is found.
- C = O stretch: For carbonyl-containing substances such as esters, ketones, and aldehydes, a strong peak occurs at approximately 1,700 cm^{-1}.
- C–H stretch: For alkanes or alkenes, peaks occur between 2,900 and 3,000 cm^{-1}.

Comparing solids versus solutions:
Solids:
- More complex sample preparation (e.g., pellets or ATR).
- Often gives information about the bulk structure (e.g., crystalline vs. amorphous).
- Can provide surface-specific information using DRIFTS.

Solutions:
- Easier sample preparation, especially for liquid phase samples.
- More attention needed to choose the right solvent to minimize interference.
- Good for studying reaction intermediates or solute–solvent interactions.

Generally, IR spectroscopy is a versatile tool for both solid and solution states, providing a wealth of information about molecular structures and interactions. Each state requires tailored techniques and considerations, but the core principle remains the same: molecules absorb specific wavelengths of IR light, producing a spectrum that can be analyzed to determine structural information.

Apparatus
- Mortal and pestel
- Hydraulic press
- Electronic balance

Instrument: FTIR spectrometer

Chemicals:
- Ethanol
- Benzene or toluene
- KBr pellet
- CCl_4
- Vanillin
- Tetrachloroethylene

Procedure
Obtaining the spectrum – FTIR
Utilize the air in the background (*). After the polystyrene film is placed in the IR beam, the IR spectra of the film can be obtained. You can calibrate the instrument using this spectrum. To guarantee that atmospheric carbon dioxide is expelled from the samples, run the proper background for each sample, place it in the IR beam, shut the cover, and collect the spectrum after two minutes. Obtain the sample's IR spectrum by first running the background on either (a) air, (b) salt plates without any sample, or (c) the fixed path length cell with the solvent, as needed. Always close the sample compartment cover and let moisture- and CO_2-free air flow through the instrument's cell compartment for approximately 2 min before beginning any data collection for the spectrum generation.

Sample preparation: solids

Depending on the region of interest in the IR spectrum, the solid substance is suspended in oil (such as Nujol, a hydrocarbon liquid, or Fluorolube 1,370–4,000 cm^{-1}, a fluorinated hydrocarbon liquid). Next, the suspension (paste) will be kept in a thin layer between two KBr salt plates that have been polished (see page 61). Using an agate mortar and pestle, grind a dry sample of the solid (glycine) for around 2–3 min until it becomes a fine powder (1–2 µm particle size < IR radiation wavelength, which would cake at the pestle's edge and take on a glossy appearance). Once a drop of the hydrocarbon oil has been added, grind the powder until it forms a fine, smooth, uniform paste in the oil.

Five more minutes of grinding is all that is needed to get the right consistency. Spoon a tiny bit of this paste onto a single salt plate; always handle these plates by the edges. To eliminate any air pockets and distribute the sample uniformly, place another plate over the "paste" and push it down. After positioning the plates in a demountable cell holder, tighten the screws to a little looser fit than with your fingers. Next, proceed to tighten the screws in a diagonal and cyclic manner until they are finger tight. Utilizing the FTIR spectrometer as outlined below, obtain the IR spectrum.

A new sample is necessary if the Nujol peaks outnumber the set of observed peaks, which indicates inadequate sample preparation, or if the peaks are wide, which indicates inadequate grounding of the solid sample. Reduce the amount of sample in the plates if the peaks are too strong. You can do this by gently pressing the plates together or by scraping off material with a clean blade. Do not use screws to compress the sample that is sandwiched; instead, reset the cell. After using, wipe the salt plates down with a paper towel dampened with CCl_4 and pat dry with a lint-free towel.

KBr pellet:
This method uses a thin wafer of a closed mixture of the sample in spectra-grade KBr. In a dry, clean agate mortar, take a 100 mg dry glycine sample and grind it very finely to reduce the size of the particles. A plastic vial should be filled with 1–5 mg of the glycine sample, 500 mg of dry spectra-grade KBr solid, and thoroughly mixed. This mixture should be transferred to a die assembly and evenly spread. Under vacuum, the sample will finally be pressed to a weight of about 10,000–15,000 pounds. For approximately 2 min, maintain this pressure on the sample. The final KBr pellet needs to be light transparent. To get the sample's IR spectrum, mount the KBr pellet containing it on a holder.

Thin-film preparation:
About 2 mL of ethanol should be used to dissolve approximately 0.01 g of the solid (vanillin). A polished NaCl crystal should have a few drops of the solution evaporated from it. Allow the ethanol to evaporate, as it is volatile. Obtain the generated thin film's IR spectrum.

Sample preparation of solutions

Mix approximately 100 μL of benzene with approximately 900 μL of solvent to create a ~10% (v/v) solution of either toluene or benzene in tetrachloroethylene. To acquire the spectrum of a fixed path length IR cell with tetrachloroethylene as the background, fill a fixed path IR cell with the solvent and seal it with Teflon stoppers. Eliminate the solvent entirely. Using a sterile hypodermic needle, add the benzene solution to the predetermined path length cell. Keep the Teflon stoppers closed. Proceed with the spectrum as you previously did.

As soon as possible, clean the cell with acetone and then with a nitrogen stream. Place it in the appropriate container and secure the lid. Obtain the neat liquid's IR spectrum as a capillary film between two NaCl plates.

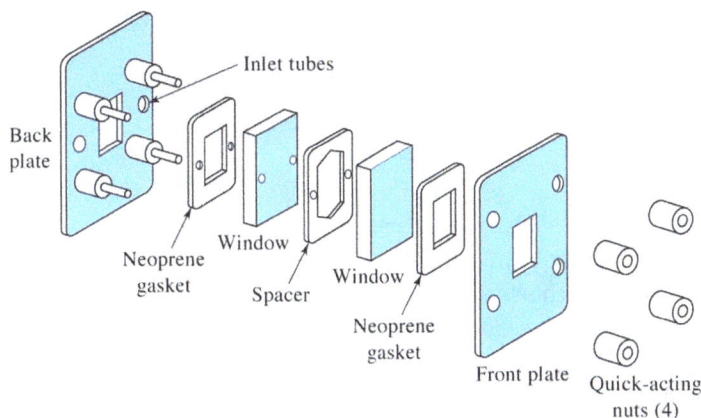

Figure 8: Infrared salt plate assembly.

Calibration of the fixed path length IR cell – interference fringe method

This experiment measures the interference fringes that result from the constructive and destructive interference of electromagnetic waves. In the %T mode, set the output. Gather the air backdrop file. After inserting the empty cell into the sample compartment, run the "IR" on a fixed path length cell (in this case, a ZnSe cell). This will produce a pattern of fringes. For linear frequency instruments such as the one we have, the relationship between the route length t, the number of peaks, n, between two wave numbers, say, v_0 (the reference peak) and the nth peak (fringe) left of reference peak, say v_n, is

$$t = \frac{10n}{2(v_n - v_0)}$$

Wavenumber is expressed in cm^{-1}, and cell spacing thickness is expressed as t (in millimeters). To create a linear plot of wavenumber versus fringe number (n), alter the above equation.

When we rearrange the preceding equation for a range of wave numbers, we obtain

$$v_n = \frac{5}{t}n + v_0$$

Choose a peak in the middle of the plot to serve as a "reference peak" on the low frequency side (let us say it is v_0); then, identify the first peak (v_1) on the high wavenumber side of the "reference peak." For the latter peak, $n = 1$ (the index) is set to 1. Proceed to determine the frequencies for a sequence of successive peaks on the high wavenumber side ($n = 2, 3, \ldots, 8$) and note the associated values of n and v_n. Plot the data of v_n against n. From the slope of the best fit line, get the cell's path length and the corresponding uncertainty in that length.

FTIR: instrument setup

The Nicolet FTIR spectrometer is left running continuously to stabilize the system. Open the OMNIC software. On the right and bottom of the OMNIC window, there will be a Bench Status indicator. It shows a green check mark indicating the system is ready to be used. If a red cross or a yellow circle shows up, it would require the attention of a service person.

Open the **Experiment Setup** window (under the **Collect menu**). Choose the **Collect tab** and check that the following settings have been entered before clicking OK:
- Number of scans: 32
- Resolution: 4
- Final format: Transmittance
- Correction: None
- File handling: Auto save
- Background handling: Collect background after 300 min

Background and sample spectrum

For the next 5 h, this spectrometer will be using the background spectrum that you are expected to acquire. If the background "cell" is not being used, that is, air is the backdrop, then leave the door closed. To open the spectrometer sample area, turn the handle in the center of the front face of the spectrometer.

Except for "cell" exchanges into the sample compartment, always keep the door to the spectrometer compartment closed. Click OK after choosing Collect Background from the Collect menu or by using the toolbar icon. The (background) spectrum will show up on the computer screen as the device gathers data. As the 32 scans are added up and averaged, a box at the bottom left of the spectrum will show you how the scans are progressing. You will be prompted to add the spectrum to window 1 after the experiment is complete; select Yes.

Using the toolbar icon or the Collect menu, choose Collect Sample, then click OK. Give it a title. You will be prompted to add the spectrum to window 1 after the experiment is complete; select Yes. Click on the background spectrum, choose Edit from the menu bar, and then choose Clear to remove it from the screen. Click File and choose Print.

Experiment 11: spectrophotometric determination of iron by 1,10-phenanthroline in drinking water

Objective: To measure the concentration of iron using a calibration curve.

Theory

It is crucial for numerous sectors to be able to quickly and effectively determine the amount of iron present in aqueous solutions. Determine the amount of iron in waste streams for manufacturing sectors that require metal parts to be cleaned in order to comply with environmental regulations. In order to ensure legal compliance, the safety of the water supply for animals and the general public, governments at all levels have an interest in testing drinking water, natural waterways, and wastewater to establish the iron content.

Iron content analysis using atomic emission spectroscopy in an inductively coupled plasma (ICP) spectrometer can be carried out in well-equipped, contemporary labs. Atomic absorption spectrometry in flame mode is another option, however iron solutions are known to clog burners with iron oxide at higher concentrations. A substantial investment in instrumentation and a continuous laboratory infrastructure including compressed gases and exhaust vapor control are necessary for both of these methods. Thankfully, the colorimetric iron content determination method's solution chemistry can be easily simplified into a kit that can be used in the field with portable equipment or in the lab with an inexpensive visible spectrophotometer and basic glassware.

Preparing the colored iron complex

The colorimetric determination of iron content involves the measurement of the ferrous ion (Fe^{2+}) when it forms a complex with three molecules of 1, 10-phenanthroline, also called *ortho*-phenanthroline or abbreviated as phen.

A commonly used method for the determination of trace amounts of iron involves the complexation of Fe^{2+} with 1,10-phenanthroline (phen) to produce an intensely red orange-colored complex. Since the iron present in the water predominantly exists as Fe^{3+}, it is necessary to first reduce Fe^{2+} to Fe^{3+}. By adding the reducing agent hydroxylamine, this is achieved. Because dissolved oxygen will reoxidize Fe^{2+} to Fe^{3+}, an excess of reducing agent is required to keep iron in the +2 state. 1,10-Phenanthroline quantitatively complexes Fe^{2+} at pH values between 3 and 9. To keep the pH at 3.5 consistently, sodium acetate is utilized as a buffer. Fe^{2+} will be converted to Fe^{3+} if the pH is too high, while H^+ will compete with Fe^{2+} for the basic 1,10-phenanthroline (forming $phenH^+$) if the pH is too low. In either case, total complexation will not occur. In the report's Introduction section, you should go over these potential issues and their implications.

https://doi.org/10.1515/9783111702261-012

With a spectrophotometer set at a fixed wavelength of 508 nm and an external calibration based on iron standard solutions, the iron-phen complex is determined. The vivid orange complex known as ferrous tris(1, 10-phenanthroline)iron(II) or [Fe(phen)3] $^{2+}$ is created when 1,10-phenanthroline and Fe^{2+} combine. Scheme 2 displays the 1,10-phenanthroline's chemical structure and numbering system.

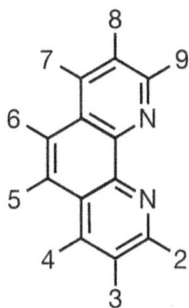

Scheme 2: Numbering scheme for positions in 1,10-phenanthroline.

Iron can exist in two different ion forms: Fe^{2+} (ferrous) and Fe^{3+} (ferric). Prior to adding the phenanthroline to form the complex, any Fe^{3+} in solution must be reduced to Fe^{2+} in order to do a total iron measurement. In this experimental approach, hydroxylamine hydrochloride is the selected reducing agent. It interacts with Fe^{3+} through the following reaction:

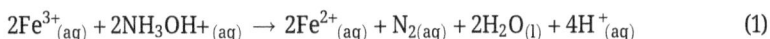

$$2Fe^{3+}_{(aq)} + 2NH_3OH+_{(aq)} \rightarrow 2Fe^{2+}_{(aq)} + N_{2(aq)} + 2H_2O_{(l)} + 4H^+_{(aq)} \tag{1}$$

Upon adding the phenanthroline, the following reaction occurs:

$$Fe^{2+} + 3phen \rightarrow \left[Fe(phen)_3 \right]^{2+} \tag{2}$$

Using Beer's law, the concentration of an unknown substance is calculated. We must first use Beer's law to calibrate the procedure with the spectrophotometer in order to ascertain the iron concentration in an unknown solution:

$$A = \varepsilon bc \qquad Beer's\,law$$

where A is the absorbance reported by the spectrophotometer; ε is the extinction coefficient, a value that describes how strongly the particular compound absorbs photons at the particular wavelength, typically with units of (L/cm·mol); b is the path length of the cuvette in cm, where typically a 1 cm path length cuvette is used; c is the concentration of the solution in mol/L (mol·L^{-1}).

One way to create a Beer's law plot is to prepare a number of solutions with known concentrations and plot each solution's absorbance on the y-axis against the concentration on the x-axis: When we compare the plot with the equation for Beer's law, we can observe that the line's slope equals ε. There are two methods we can use the plot to determine the concentration of an unknown solution:

1. To determine the point on the y-axis that represents the measured absorbance, use the plot itself. Draw a vertical line directly down to the concentration on the x-axis after tracing a horizontal line from that point to the plot line. The concentration of the unidentified solution is represented by the point where this touches the x-axis.

2. Apply the line's equation. $C = A/\varepsilon$ if $A = \varepsilon bc$ and $b = 1$. Plot your data using a spreadsheet application or a graphing calculator, then identify the best-fit line (trend line) to get the slope of your line. This angle comes to ε. To find the concentration of the unknown solution, divide the measured A value for the unknown by ε.

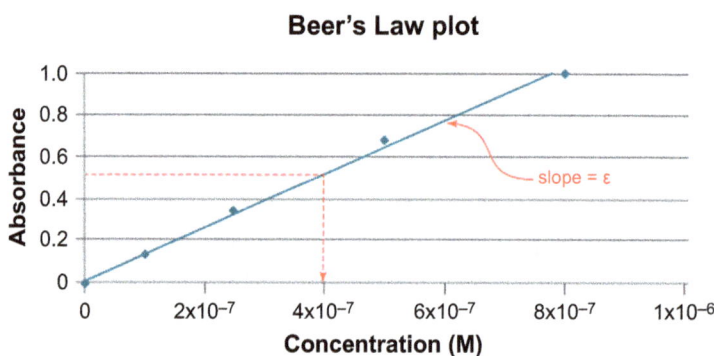

Figure 9: Example of a generic Beer's law plot.

Apparatus
- Plastic cuvette
- Volumetric flask
- 500 µL automatic pipettor
- Pipette

Chemicals
- Ammonium iron(II) sulfate, hexahydrate
- Hydroxylamine hydrochloride
- 1,10-Phenanthroline
- Sodium acetate

Procedure
1. Switch on the device and give time to finish the startup process. Give the instrument at least 30 min to warm up and stable. In the spectrophotometer program, set up the desired experiment. To use in your investigations, get a glass test tube

or a square plastic cuvette. If the test tube cuvette is not already marked with a white line, mark it close to the top with a pen. You can make sure that the mark is placed consistently inside the instrument.

2. Fill the cuvette with liquid until there is around 3 cm (or 4 cm in test tube cases) in the bottom. Use a plastic transfer pipette if one is available.
    As long as the liquid level in the cuvette is more than 3 cm, precise measurements are not important. Avoid wasting solution or running the danger of leaks by not filling the cuvette all the way.

3. Put the cuvette into the SPECTRONIC 200 visible spectrophotometer's sample stage. The transparent sides of a plastic cuvette should be on the left and right when utilizing one. Position the test tube cuvette with the mark facing toward the right if you are using one.

4. Use water or the suitable "blank" solvent to carry out procedures 2 and 3 after the warm-up period. To zero the instrument, push the autozero button.

5. Shake out as much of the rinse solvent as you can after emptying and rinsing the cuvette for each subsequent measurement.

Never put extra solution back in the stock bottle while making samples. Transfer all rubbish or surplus into the designated waste receptacle. Use your sample as a guide for stages 2 and 3.

Part 1: prepare the reagents and standards

1. Before starting this experiment, prepare the solutions. Use deionized water for all dilutions.
    -   Iron solution (100 mg/L): In a 1 L volumetric flask, dissolve 0.7022 g of ammonium iron(II) sulfate, hexahydrate, in water.
    -   Iron working solution (10 mg/L): A 50 mL volumetric flask should be filled to the brim with water after 5 mL of the 100 mg/L iron solution has been pipetted into it.
    -   A solution of 0.3 M hydroxylamine hydrochloride
    -   0.25% 1,10-phenanthroline solution: Use heat if needed to stir the mixture and make sure all of the solids have dissolved.
    -   A sodium acetate solution at 1.0 M

2. Set up six 50 mL volumetric flasks and pipette reagents into them as follows:

Flask number	10 mg/L iron solution (mL)	0.3 M hydroxylamine hydrochloride solution (mL)
1	0	1.00
2	2	1.00
3	5	1.00
4	8	1.00
5	14	1.00
6	20	1.00

Stopper each flask, then invert repeatedly for 2 min to allow the reaction to complete.
3. To each flask, add:
- 5.0 mL of 1.0 M sodium acetate solution. Stopper and invert the flask to mix.
- 5.0 mL of 0.25% 1,10-phenanthroline solution. Stopper and invert the flask to mix.
- Add deionized water to the mark. Stopper and invert several times to mix.

Part 2: determine the proper analytical wavelength
1. Transfer 3 mL of flask 1's solution into a cuvette. This will be your go-to blank or the blank will be deionized water in order to detect any potential absorbance from the matrix (i.e., the other reagents). After repeatedly rinsing the cuvette with tap water and then deionized water, fill it with the latter, set it in the holder, and then blank the spectrometer.
2. Using a lab tissue, wipe the cuvette's exterior faces. Then, insert it into the square cuvette stage of the SPECTRONIC 200 spectrophotometer sample compartment, making that the clear faces face left and right.
3. Shut off the SPECTRONIC 200 spectrophotometer's cover. To record a baseline, set up an ABS mode scan from 400 nm to 900 nm and press the 0.00 button. Hold off until the hourglass icon vanishes, signifying that the scan is finished.
4. Lift the cover, take out the cuvette containing the blank solution, and place it somewhere else. If you just have one cuvette, rinse it with deionized water before reusing it and throw away the blank solution.
5. Transfer 3 mL of flask 6's solution into a cuvette. The cuvette should be placed in the square cuvette stage of the SPECTRONIC 200 spectrophotometer sample compartment with its clear faces pointed to the left and right after being cleaned on the outside with a lab tissue. To capture the scan, close the lid and hit the circular Enter button.
6. Use the λ knob to move the green pointer line to the peak's highest place when the scan displays on the screen. You can also use the left and right arrow keys.
   Note down the λ_{max}, or wavelength of maximum absorption.
7. If the printer on your SPECTRONIC 200 spectrophotometer is working, print the screen.

Part 3: prepare the Beer's law plot
1. Determine the iron concentrations in flasks 1 through 6 and record the results in the lab report.
2. Utilize the Live Display Mode on the SPECTRONIC 200 spectrophotometer and record data in absorbance at λ_{max}, as established in Part 2.
3. To record a blank value, fill, wipe, and orient a cuvette as instructed in Part 2 using a cuvette filled with blank solution from flask 1.

4. Assemble cuvettes with the solutions from flasks 2 through 6 and measure their absorbance. In your lab report, note the values in data table 1.
5. Use your data to plot Beer's law. Calculate the line's slope, and then enter this value in your lab report as ε.

Part 4: determine the iron concentration in an unknown sample
1. Make sure one of the volumetric flasks is clean.
2. Fill the flask to the brim with 5.0 mL of an unidentified iron solution. The process for getting the solution ready for measurement is the same as in Part 1, steps 2 and 3. Take note of the solution's absorbance at λ_{max} using the SPECTRONIC 200 spectrophotometer.
3. Using the Beer's law equation and the number you computed for ε, determine the mathematical concentration of iron in the solution.
4. When calculating the unknown's concentration, don't forget to account for the impact of dilution, from 5 to 50 mL, prior to measurement.

Results

Flask number	Fe^{2+} concentration	Measured absorbance
1		
2		
3		
4		
5		
6		

Report: In preparing the partial report for this part of the experiment, you should consider/complete/discuss the following:
(1) Determine the iron's analytical concentration in the reference iron solutions. Total the observed absorbance and the calculated iron concentrations (in ppm and molarity) for each standard, sample of tap water, and unidentified sample. (Remember to factor in the dilution factors while performing your calculations.) Based on the challenge unknowns, remark on the recovery and determination's accuracy. Based on the standard error of the intercept, determine the method's detection and quantitation limitations. How should the concentration of any unknowns that go below these bounds be reported?
(2) Plot the standards' absorbance against iron concentration (including that of the iron-free solution). To find the Beer's law equation, which is expressed as $A = m[Fe] + b$, use the least-squares method, also known as linear regression. Make sure the report's Results and Abstract sections contain the calibration equation. Provide an explanation for every nonzero value of the intercept along with a note on the R^2 value.

Experiment 12: absorption spectrum of an indicator in dependence on the pH value

Objective: To determine the absorption spectrum of an indicator in dependence on the pH value.

Theory

By adding an acid/base indicator of known K_a and measuring the relative concentrations of the acid and base forms of the indicator using spectrophotometry, one can ascertain the pH of an unknown solution. The pH of seawater has been continuously monitored on board ships using this technique. The equilibrium describes the interaction between the two forms of the indicator in an aqueous solution. The absorption spectrum of a pH indicator changes as a function of pH due to the protonation and deprotonation of the indicator molecules. Here is a detailed overview of how this works, along with some practical examples.

Key concepts

1. **Protonation and deprotonation:**
    Indicators are typically weak acids or bases that exist in two forms: protonated (HIn) and deprotonated (In⁻).
    The protonated form often has a different color and absorption spectrum than the deprotonated form.

2. **pH dependence:**
    At different pH levels, the ratio of HIn to In⁻ changes, leading to variations in the absorption properties.
    The pH at which the color change occurs corresponds to the pK_a of the indicator.

3. **Absorption spectrum:**
    The spectrum displays how much light the substance absorbs at various wavelengths:
    - Peaks in the spectrum indicate the wavelengths at which the indicator absorbs light, and these can shift based on the pH.

Example indicators

1. **Phenolphthalein:**
    pH < 8.2: Colorless (HIn).
    pH > 10: Pink (In⁻).

Absorption spectrum: The spectrum shows a peak around 550 nm in the deprotonated form, while the protonated form has minimal absorbance in this region.

https://doi.org/10.1515/9783111702261-013

2. **Methyl orange**:
   pH < 3.1: Red (HIn).
   pH > 4.4: Yellow (In⁻).

Absorption spectrum: The red form absorbs strongly at around 520 nm, while the yellow form has a peak around 430 nm.

Data analysis

Peak shifts: Examine the movement of the absorption peaks as the pH changes. This change aids in determining the indicator's pK_a.

Correlate the spectral data with the indicator's visible color changes using the color change correlation.

Establish pK_a: Using the data, ascertain the pK_a, or the point at which the concentrations of HIn and In⁻ are equal.

The absorbance of a mixture at a certain wavelength is determined by the sum of the absorbances of its constituent parts. When measuring the absorbance of two overlapping components, two wavelengths must be employed:

$$A_{\lambda_1} = C_1(\mathcal{E}_1)\ \lambda_1 + C_2(\mathcal{E}_2)\ \lambda_2$$
$$A_{\lambda_2} = C_1(\mathcal{E}_1)\ \lambda_1 + C_2(\mathcal{E}_2)_{\lambda_2}$$

(1)

where the subscripts 1 and 2 indicate the two wavelengths. It is necessary to understand each form's molar absorptivity in order to calculate the amount of each form in a mixture based on the measured absorbances. The absorption spectra of the indicator's acid and basic versions are determined independently in an initial experiment. Every constituent's maximum absorbance wavelength is found, and each component's molar absorptivities at these two wavelengths are calculated. The absorbance of the solution with an unknown pH is measured at the same two wavelengths, and by solving the two simultaneous equations describing the solution absorbances at the two wavelengths, the concentrations of the two forms of the indicator in this solution are determined.

Simple weak acids (or bases) that display distinct colors in solution based on whether they are in their basic form (In⁻) or their acidic form (HIn) are known as acid–base indicators. The equilibrium indicated below will be forced either toward reactants (Hin) or products (In⁻) as the pH of a solution containing the indicator changes, causing the color of the solution to change based on the concentration of each form present. For instance, the majority of the indicator in a very acidic solution will take the form of HIn, resulting in a solution hue that matches HIn's. Strongly simple solutions will have color 2 predominating since the majority of the indicator will be in the Inform.

Depending on the proportions of HIn and In⁻ present, the solution color at intermediate pH levels will be a mixture of colors 1 and 2:

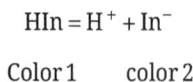

$$HIn = H^+ + In^-$$

Color 1   color 2

The indicator's acid dissociation constant (K_a) quantitatively describes the relationship between the pH (or more accurately, the hydronium ion concentration) and the relative concentrations of HIn and In⁻ in solution. This relationship is illustrated in eq. (2), where the square-bracketed terms denote the molar concentrations of each species in solution (M):

$$K_a = \frac{[H^+][In^-]}{[HIn]} \tag{2}$$

Since we usually test a solution's pH rather than its [H⁺] in practice, it helps to make a few mathematical adjustments to the K_a formula to put things in terms of quantities that have been determined by experimentation. The following is what we get when we first take the negative logarithm of both sides of the equation:

$$-\log K_a = -\log [H^+] - \log\left(\frac{[In^-]}{[HIn]}\right) \tag{3}$$

That equation can be rewritten as

$$pK_a = pH - \log\left(\frac{[In^-]}{[HIn]}\right) \tag{4}$$

Based on a closer look at Equation 4, we should be able to find the K_a for the indicator (or any weak acid, for that matter) if we can keep an eye on the relative concentrations of HIn and In⁻. Determining the relative amounts of the two forms of indicator takes a little more work than simply using a glass electrode to monitor the pH of a solution. The method employed in this experiment makes use of the indicator's ability to absorb more strongly at various wavelengths in its basic (In⁻) and acidic (Hin⁻) forms. Think about the generic indicator Hin.

The indicator is entirely in the HIn form and the absorbance that is caused by HIn is at its highest at low pH (Figure 10A). Similarly, at high pH, the absorbance resulting from In⁻ is at its highest and the indicator is fully in the In⁻ form (Figure 10B). The solution has noticeable quantities of both HIn and In⁻ at an intermediate pH, and its absorbance spectrum shows contributions from both forms. Remember Beer's law? As pH varies, the relative concentrations of each form will alter based on the indicator's K_a, which will alter each form's subsequent absorbance.

An indicator's K_a value can be determined by tracking how each form's absorbance changes in relation to the pH of the solution. When the concentrations of HIn and In⁻ in solution are equal, eq. (4) states that the pH of the solution will equal the

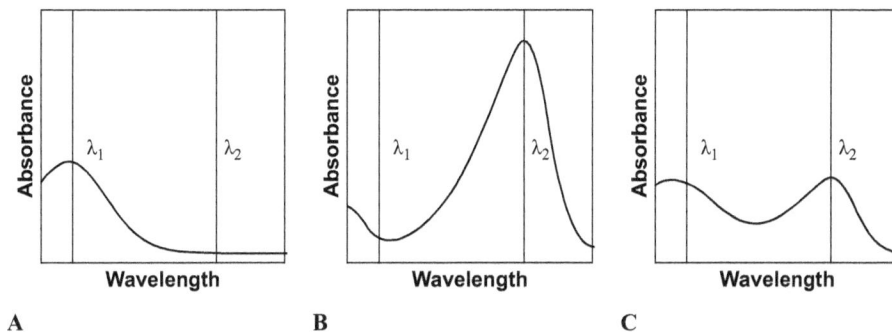

Figure 10: Example absorbance spectra of an acid–base indicator in (A) acidic solution, (B) basic solution, and (C) solution of intermediate pH.

indicator's pK_a. According to the experiments, this is equivalent to a pH at which each form's absorbance is half of its maximal absorbance. By now, half of the original HIn moles had been changed into an equal number of In^- moles. Consequently, the log component in eq. (4) becomes zero and the concentrations of each species are equal.

Therefore, as illustrated in Figure 11, the pK_a of an indicator correlates to the pH of the solution at the inflection point in a plot of absorbance as a function of pH. Nonetheless, keep in mind that it is crucial that the absorbance at each wavelength matches the absorbance of only In^- at λ_2 and only HIn at λ_1.

The Henderson–Hasselbach equation for the indicator takes on its most popular version when eq. (4) is rearranged. An alternative method for calculating an indicator's pK_a is given by this expression, which involves graphing pH (y-axis) as a function of $\log([In^-]/[HIn])$. When pH equals pK_a, as shown in Figure 12, the resulting straight line crosses the pH axis:

$$pH = pK_a + \log\left(\frac{[In^-]}{[HIn]}\right) \tag{5}$$

The ratio $[In^-]/[HIn]$, which is needed for the plot, can be found by two wavelength spectrophotometric measurements. The indicator's acidic form (Hin) absorbs light more strongly than its basic form (In^-), which is why the first wavelength (λ_1) is selected. The second wavelength (λ_2) is selected because radiation is extensively absorbed by the basic form but not by the acidic form. The absorbance at λ_1 and λ_2, assuming adherence to Beer's law, is:

$$A_{\lambda_1} = \varepsilon(HIn, \lambda_1) \times b \times [HIn] \tag{6}$$

$$A_{\lambda_2} = \varepsilon(In^-, \lambda_2) \times b \times [In^-] \tag{7}$$

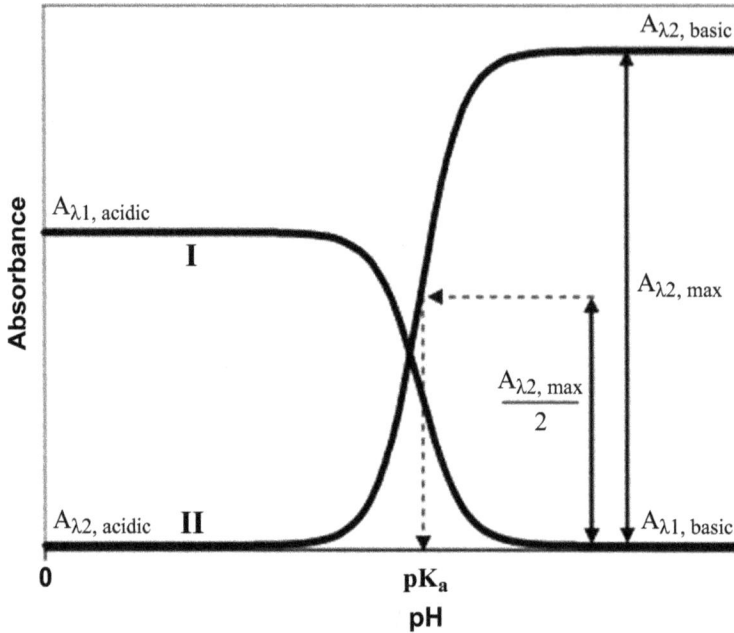

Figure 11: A plot of absorbance as a function of pH for an indicator. I – measurements that are made at a wavelength (λ_1) at which HIn absorbs radiation; II – measurements that are made at a wavelength (λ_2) at which In⁻ absorbs radiation.

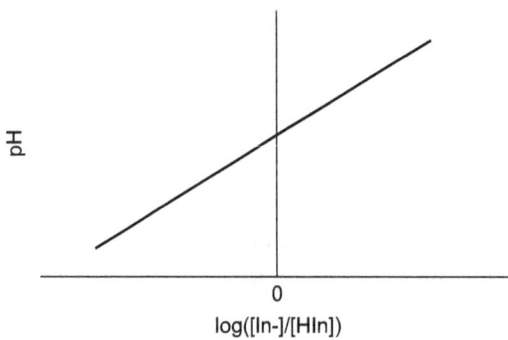

Figure 12: A plot of pH as a function of log([In⁻]/[HIn]). The pK_a of the indicator corresponds to the intersection of the line with the pH axis.

where A is the absorbance, ε is the molar absorptivity, and b is the cell path length. At any pH, the total concentration (C_T) of both forms of the indicator is constant and the sum of the individual concentrations of each species:

$$C_T = [HIn] + [In^-] \tag{8}$$

Nearly majority of the indicator is in the acidic form in low pH solutions. Thus, in extremely acidic solutions, $C_T = [HIn]$, and

$$A_{\lambda_1,\text{acidic}} = \mathcal{E}(HIn, \lambda_1) \times b \times C_T \tag{9}$$

Almost all of the indication in extremely basic solutions is in the basic form, and $C_T = [In^-]$ so that

$$A_{\lambda_2,\text{basic}} = \mathcal{E}(In^-, \lambda_2) \times b \times C_T \tag{10}$$

Taking the ratio of equations (6)–(9) and (7)–(10) provides expressions for the fractions of the indicator present in each form:

$$\frac{A_{\lambda_1}}{A_{\lambda_1,\text{acidic}}} = \frac{\varepsilon\,In*b[HIn]}{\varepsilon\,In*bC_T} = \frac{[HIn]}{C_T} \tag{11}$$

$$\frac{A_{\lambda_2}}{A_{\lambda_2,\text{basic}}} = \frac{\varepsilon\,In*b[In]}{\varepsilon\,In*bC_T} = \frac{[In]}{C_T} \tag{12}$$

The ratio $[In^-]/[HIn]$ at any pH can be obtained by dividing eq. (11) by eq. (12) to give

$$\frac{[In]}{[HIn]} = \frac{A_{\lambda_2}/A_{\lambda_2,\text{basic}}}{A_{\lambda_1}/A_{\lambda_1,\text{acidic}}} = \frac{A_{\lambda_2}*A_{\lambda_1,\text{acidic}}}{A_{\lambda_1}*A_{\lambda_2,\text{basic}}} \tag{13}$$

With background corrected absorbances, the latter equation can be used to calculate $[In^-]/[HIn]$ from absorbance data obtained at any pH. This adjustment takes into consideration the absorbance at λ_1 and λ_2 that is contributed by other species' absorbance in solution. For instance, at λ_2, we presume that the measured absorbance solely stems from the indicator's basic form. Indeed, considerable absorption may occur at λ_2 for the acidic form (or other species in solution), and at λ_1 for the basic form. To account for this, deduct from each of the other values performed at that wavelength the minimum (or background) absorbance observed at λ_2. For example,

$$A_{\lambda_2} = A_{\lambda_2}(\text{measured}) - A_{\lambda_2}(\text{min}) \tag{14}$$

Your most acidic solution's measurement at λ_2 should yield the result $A_{\lambda2}$ (min). Using the two previously stated approaches, the pK_a of bromothymol blue (3',3"-dibromothymolsulfonephthalein) is found in the experiment. The indicator is yellow for pH values below 6, and blue at pH values above 7.6. The blue and yellow mix to form a green solution at an intermediate pH.

Apparatus
- Cuvettes
- Graduated cylinder
- pH meter and electrodes
- Pipette
- Volumetric flasks

Instrument
Scanning spectrophotometer

Chemicals
- Bromothymol blue solution (0.1% in 20% ethanol)
- Hydrochloric acid (concentrated)
- KH_2PO_4 solution (0.10 M)
- Na_2HPO_4 solution (0.10 M)
- Sodium hydroxide solution (4 M)

Procedure
1. Fill two labeled 25 mL volumetric flasks with 1.00 mL of the bromothymol blue solution each using a pipette. To one of the flasks, add 4 drops of strong hydrochloric acid and 5 mL of distilled or deionized water. Adjust the solution's concentration with water. The pH of the final mixture should be roughly 1. Add 12 drops of the 4 M sodium hydroxide solution to the second flask, then top out the flask with water. The pH of the mixture need to be roughly 13.
2. Place numbers 1 through 10 on the labels of ten 25 mL volumetric flasks. About 1.00 mL of the bromothymol blue solution should be added to each flask using a pipette.
    To each flask, add the volumes of 0.10 M KH_2PO_4 and 0.10 M Na_2HPO_4 solutions that are shown in the table. Adjust the strength of every solution with water. In this stage, buffer solutions with various pH values are made.

Flask number	KH_2O_4 (mL)	Na_2HPO_4 (mL)
	5	0
	10	0.5
	5	1
	10	5
	5	5
	5	10
	1	5
	1	10
	05	10
	0	5

3. Check the pH values of pH 1, hydrochloric acid solution, pH 13, sodium hydroxide solution, and the 10 solutions that were made in this step using the pH meter. Note down each solution's pH.
4. Take and record the pH 1 and pH 13 bromothymol blue solutions' spectra between 300 and 800 nm using the scanning spectrophotometer. You will learn how to play the instrument from your tutor. Select one wavelength (λ_1), at which the pH 1 solution absorbs strongly, and a second wavelength (λ_2), at which the pH 13 solution absorbs strongly, but the pH 1 solution absorbs poorly, from the recorded spectra. Keep in mind that low % transmittance is correlated with high absorption.
5. Using water as a blank, measure and tabulate the absorbance of each of the 12 solutions at the two selected wavelengths. Printing a spectrum for the buffered solutions is not required.

Worksheets
1. Create a plot of the absorbance (*y*-axis) for each of the 12 solutions at each of the two wavelengths versus the pH (*x*-axis). Determine the indicator's pK_a value from each plot's inflection point by joining the points with a smooth line.
2. Calculate the minimum absorbance at each wavelength from each of the two plots. Except in cases where an error was made in one of the measurements, the minimum absorbance corresponds to that of the pH 1 solution for 2 and to that of the pH 13 solution for 1.
3. Take the absorbance of each of the 12 solutions at the wavelength and subtract it from the minimum absorbance at that wavelength. After background absorbance correction, the resultant absorbances are calculated.
4. Calculate the adjusted absorbance at λ_1 (A λ_1, acidic) and λ_2 (A λ_2, basic) in the acidic and basic solutions, respectively. According to the figure, these absorbances are the adjusted absorbances that match the level area at the top of each plot.
5. Determine the ratio [In$^-$]/[Hin] for each of the 12 pH values using eq. (13) and the adjusted absorbances.
6. Draw a graph of pH (*y*-axis) versus log ([In$^-$]/[Hin]). At the point where the line intersects the pH axis, find the bromothymol blue pK_a by drawing the best possible straight line between the data points.

Experiment 13: identification of a chemical compound using thin-layer chromatography

Objective: To perform the separation and analysis of a compound in a plant pigment using thin-layer chromatography (TLC).

Theory

In chemistry, TLC is a very useful analytical method. It allows for the quick separation of compounds, revealing the quantity and makeup of a mixture's constituent parts. TLC can also be used to monitor the advancement of a reaction, an extraction, or a purification process, identify compounds by comparing with the existing samples, and assess the compound's purity. TLC is a valuable technique for identifying and separating chemical compounds.

You will learn about the chemical principles and mechanics of TLC with this experiment. You will separate the soluble components of the spinach extract in the first section, and you will evaluate the compounds you extracted in the previous lab in the second.

Principles of TLC:

The most popular solids for TLC are alumina (Al_2O_3) and silica (SiO_2), which are applied in thin layers to a small glass or plastic plate. This is the stationary phase. An organic solvent or a combination of solvents serves as the mobile phase. A little patch of the sample combination is deposited near the bottom of the plate, and it is then placed in a jar with a few milliliters of solvent inside. Through capillary action, the solvent ascents the plate, bringing the sample mixture with it. Depending on how well it is absorbed into the stationary phase and how soluble it is in the mobile phase, each component in the mixture moves at a different pace.

The mixture's constituent parts are left behind at different distances from the point of origin as the solvent is allowed to evaporate as it approaches the top of the plate. The retention factor, or R_f value, is the ratio of the distance a chemical moves to the distance the solvent moves. This value is indicative of the stationary phase, the solvent, and the substance. In column chromatography, the mixture's individual components are separated when they elute from (leave) the column, which is made of silica or alumina. The sample is transported down the column by solvent. Although it is sluggish, this can be accomplished by allowing the solvent flow under the influence of gravity.

Using a method known as "flash chromatography," organic chemists today force the solvent through the column with a little amount of air pressure. High-pressure liquid chromatography is a comparable method for particularly challenging separations. It achieves separations by applying high solvent pressure and a very high-quality stationary phase. Relatively polar stationary phases are alumina and silica. The OH groups that are present on both sides interact significantly with polar mole-

https://doi.org/10.1515/9783111702261-014

cules. These substances travel across the plate more slowly because they are highly adsorbed, whereas nonpolar substances are only weakly absorbed and move more swiftly.

Naturally, solvent polarity has an impact on the velocity at which chemicals move. In contrast to nonpolar solvents, which move nonpolar compounds slowly or not at all, polar solvents move polar compounds swiftly. Nonpolar chemicals tend to flow more swiftly in most solvents because they do not stick to the silica as firmly.

Here is a step-by-step guide to use TLC for identification.

Materials needed:
- **TLC plates** (usually silica gel or alumina)
- **Developing solvent** (a suitable solvent or solvent mixture)
- **Sample solutions** (the compounds to be analyzed)
- **Capillary tubes** or a micropipette
- **TLC chamber** (for developing)
- **UV light source** (optional, for visualizing spots)
- **Marker** (to draw baseline)

Working procedure of TLC

TLC plate preparation:
Using a pencil, draw a horizontal line (baseline) approximately 1 cm from the bottom of the TLC plate. For every sample you wish to examine, make small marks along the baseline.

Sample spotting: Apply small quantities of each sample solution to the designated areas on the baseline using a capillary tube. Before continuing, allow the areas to dry.

Setting up the developing chamber: Fill the TLC chamber with a small quantity of the developing solvent, making that the solvent level is lower than the baseline.

Plate development: Gently insert the TLC plate into the chamber and cover it.
Do not allow the solvent to reach the top edge; instead, let it ascend the plate via capillary action until it approaches the top.

Plate removal: After the solvent front has risen to the appropriate height, remove the plate from the chamber. Mark the solvent front with a pencil right away.

Visualizing the spots: If the chemicals are not visible, visualize the spots using a staining agent or UV light.

Calculate the separation between each location and the baseline.

Calculating R_f values:
Calculate the retention factor (R_f) for each compound using the formula:

$$R_f = \frac{\text{Distance travelled by the component}}{\text{Distance travelled by the solvent}}$$

Comparison and identification: Examine the appearance of spots and their R_f values in relation to those of established standards conducted under identical circumstances. If more identification is required, employ additional analytical methods (such as mass spectrometry, NMR, or IR). Advice for success: To guarantee reproducibility, employ consistent spotting methods.

   To ensure accuracy, run several trials.
– Be sure to pick a solvent that will work well for separating your components. A number of popular chromatographic solvents are included in Table 4 arranged in the increasing order of bulk polarity, or dielectric constant, or ε. This is an approximate eluotropic series because a solvent's chromatographic "eluting power" – or its capacity to transport compounds – is generally correlated with its polarity.

Table 4: Eluotropic series of organic solvents.

Solvent	$\varepsilon *$ (dielectric constant (debyes))
Alkanes	2
Isopropyl alcohol	18.3
Benzene	2.3
Acetone	20.7
Diethyl ether	4.3
Ethanol	24.3
Diethyl ether	4.3
Chloroform	4.7
Methanol	32.6
Ethyl acetate	6.0
Acetonitrile	37.0
Dichloromethane	8.9
Water	78.5

Apparatus:
– TLC plate
– UV lamp
– Forceps
– Chamber with lid
– Ruler and pencil
– Measuring cylinder, separatory funnel, filter paper, and capillary tube

Chemicals:
– Acetone
– Petroleum ether

- Hexane
- Ethanol

Experiment A
Pigments found in plants
Plants employ a variety of pigments in their mechanisms for gathering light. These substances are classified as carotenoids and chlorophylls. Members of these categories that are representative are β-carotene and chlorophyll (a and b). These pigments can be separated in a sample of spinach extract using TLC. Spots from various carotenes, such as β-carotene and α-carotene, which have endocyclic double bonds that are moved one position (out of conjugation) in relation to the β-isomer, as well as other oxygen-containing carotene derivatives known as xanthophylls, should be visible.

On the TLC plate, each should show up as yellow or orange spots. Furthermore, specks representing the green chlorophylls a and b and gray spots representing the pheophytins a and b should be visible. Pheophytins are simply chlorophylls with two H^+ ions in place of the magnesium ion (Mg^{2+}).

Procedure
1. To start, go to Valentine and get a medium-sized wad of fresh spinach. Just kidding. There will be a spinach extract solution available in a 1:1 acetone and petroleum ether ratio.
2. Regards. (To produce this solution, the spinach was combined with the solvent mixture and sand, and then ground into a fine paste using a mortar and pestle. The organic compounds dissolve in the solvent as a result of the sand's destruction of the cell walls. The dark green solution was then dried, filtered, and kept in a cold, dark area (a refrigerator) after being cleaned with water using a separatory funnel. Assumedly, the light goes out when the door is closed.
3. There will be supplied silica TLC plates with plastic backing, measuring 2.5 × 7.5 cm). Take caution when handling these. One plate's bottom should be the location of a light pencil line drawn from 1/3 to 1/2 inch.
4. To ensure that your sample combination is above the solvent level, estimate rather than measure; all you are doing is marking the beginning point.
5. Make three "tick" marks with the pencil, evenly spaced apart, to indicate where you will position the three different spots that you will be making along the pencil line using a capillary micropipette. (Practice on some filter paper first; attempt to make the smallest possible spot before you spot a genuine TLC plate.) Now that you are proficient, go ahead and spot the TLC plate. Make the first spot as small as possible (1 mm in diameter or less).
6. Next, press the capillary up against the plate to create a wide area. Third, to increase the concentration without unduly extending the spot, create as small a spot as you can, let the solvent to evaporate for a few seconds, then spot again.

7. Create the plate using a 1:1 hexane to ethanol combination. In this instance, speed takes precedence above accuracy.
8. Simply pipette 20 mL of each into a designated developing chamber, cover it, and give it a gentle shake to ensure that the air inside is saturated with solvent vapor.
    Use forceps to carefully insert the TLC plate, cap the chamber, and allow the solvent to rise until it gets close to the top of the plate. Be careful not to disturb the chamber.
9. Use forceps to remove the plate, then mark the solvent front's location with a pencil and let the solvent evaporate. Try it again if something went very wrong (compounds all ran to the edge, for example). Use less if you cannot see the spots well; if everything blended into one large smear in every instance, you might have spotted too much. Consult your instructor or a teaching assistant when needed.
10. Draw a circle around every area that is visible (in case light and air exposure cause them to vanish). Draw a sketch of the plate in your notebook, noting the different colors for each location.
11. Next, use one of the handheld UV lamps to expose the plate to UV radiation with a wavelength of 254 nm. Be careful: UV rays can damage your eyes. Wear goggles since they will block UV rays and avoid staring straight at the light. When exposed to light with a wavelength of 254 nm, a fluorescent indicator included in the silica TLC plates will glow green. Numerous substances have the ability to quench fluorescence, reducing its intensity and making dark patches stand out against the brilliant background. Additionally, when exposed to UV light, some spots may glow and appear bright.
12. Draw a circle around any newly appearing spots and record whether the substances are fluorescent (bright) or fluorescence quenchers (dark).
13. Then, place the plates in the lab's designated chamber and let them be exposed to I_2 vapor for a few minutes. Make a note of any newly appearing spots.
14. Instead of throwing away your used TLC plates in the trash, dispose of them in the provided waste bottle! After cutting one dish in half lengthwise with scissors, evenly distribute the spinach on each half (use one or more spots, depending on what worked best the previous time).
15. Elute one with a mixture of 1:1 and the other with 3:1 ethanol/hexane. Recall to utilize Dispo Pipettes, estimate, be quick and filthy, and avoid playing around with grad cylinders. Of course, use two different jars, but run them both simultaneously. While they are running, write the R_f values of chlorophylls, pheophytins, and carotenes/xanthophylls that you can identify based on spot color in your notebook, along with the values of any other unknown spots that appeared. This will serve as a summary of the first TLC results.

References

R. D. Braun, Introduction to Chemical Analysis, McGraw-Hill, New York, 1982, pp. 197–199.

E. B. Sandell, Colorimetric Determination of Traces of Metals, 3rd ed. Interscience Publishers, Inc., New York, 1959. https://www.stellarnet.us/wpcontent/uploads/StellarNet Quantification of Riboflavin in Energy Drinks – Exp3_Fluor_Riboflavin.pdf (Accessed July 12, 2020)

A. S. Chandami, P. P. Choudhari, M. P. Wadekar, Determination of refractive index, density, molar refraction and polarizability constant of substituted N,N'-bis (salicyliden)-arylmethanediamines in different binary mixture refractometrically.

D. A. Skoog, D. M. West, F. J. Holler, S. R. Crouch, Analytical Chemistry: An Introduction, 7th ed., Chapter 23, pp. 594–563.

A. H. Beckett, J. B. Stenlake, Practical Pharmaceutical Chemistry. In K. G. Patel, P. A. Shah, H. G. Raval, D. A. Shah & S. L. Baldania (Eds.). Laboratory Handbook in Instrumental Analysis, 1st ed, 4th ed, BS Publication Nirav Prakashan, 327.

D. A. Skoog, D. M. West, F. J. Holler, S. R. Crouch, Analytical Chemistry: An Introduction, 7th ed. Chapter 23, pp. 594–631.

(M.Pharm), D.Z.G.K. (no date) 'Lab Manual', *H. R. Patel Institute of Pharmaceutical Education and Research, Shirpur.*

Manual, T.S.U. 222 L, 'Spectrophotometric Determination of the pKa of Bromothymol Blue', no date, (4).

Singh, D. S. B., 'Spectroscopic determination of iron by 1,10-phenanthroline method', (April), 2020, pp. 1–5. Available at: https://doi.org/10.13140/RG.2.2.12030.74568.

https://doi.org/10.1515/9783111702261-015

Index

https://doi.org/10.1515/9783111702261-016

www.ingramcontent.com/pod-product-compliance
Lightning Source LLC
Chambersburg PA
CBHW081552220326
41598CB00036B/6647